Mutation Breeding in Oil Palm:

A Manual

Techniques in Plantation Science Series

Series editors:

Brian P. Forster, Lead Scientist, Verdant Bioscience, Indonesia
Peter D.S. Caligari, Science Strategy Executive Director, Verdant Bioscience, Indonesia

About the series:

A series of manuals covering techniques in plantation science that form the essential underlying needs to carry out plantation science.

The series reflects the expertise in Verdant Bioscience that underlies the plantation science activities carried out at the Verdant Plantation Science Centre at Timbang Deli, Deli Serdang, North Sumatra, Indonesia.

Titles available:

1. **Crossing in Oil Palm: A Manual** – Umi Setiawati, Baihaqi Sitepu, Fazrin Nur, Brian P. Forster and Sylvester Dery
2. **Seed Production in Oil Palm: A Manual** – Eddy S. Kelanaputra, Stephen P.C. Nelson, Umi Setiawati, Baihaqi Sitepu, Fazrin Nur, Brian P. Forster and Abdul R. Purba
3. **Nursery Screening for *Ganoderma* Response in Oil Palm Seedlings: A Manual** – Miranti Rahmaningsih, Ike Virdiana, Syamsul Bahri, Yassier Anwar, Brian P. Forster and Frédéric Breton
4. **Mutation Breeding in Oil Palm: A Manual** – Fazrin Nur, Brian P. Forster, Samuel A. Osei, Samuel Amiteye, Jennifer Ciomas, Soeranto Hoeman and Ljupcho Jankuloski

Mutation Breeding in Oil Palm:
A Manual

Fazrin Nur
Verdant Bioscience, Indonesia

Brian P. Forster
Verdant Bioscience, Indonesia

Samuel A. Osei
Oil Palm Research Institute, Ghana

Samuel Amiteye
Ghana Atomic Energy Commission, Ghana

Jennifer Ciomas
Verdant Bioscience, Indonesia

Soeranto Hoeman
National Nuclear Energy Agency, Indonesia

Ljupcho Jankuloski
Joint FAO/IAEA Division of Nuclear Techniques in Food and Agriculture, Austria

CABI is a trading name of CAB International

CABI	CABI
Nosworthy Way	745 Atlantic Avenue
Wallingford	8th Floor
Oxfordshire OX10 8DE	Boston, MA 02111
UK	USA
Tel: +44 (0)1491 832111	Tel: +1 (617)682-9015
Fax: +44 (0)1491 833508	E-mail: cabi-nao@cabi.org
E-mail: info@cabi.org	
Website: www.cabi.org	

A catalogue record for this book is available from the British Library, London, UK.

Library of Congress Cataloging-in-Publication Data

Names: Nur, Fazrin, author.
Title: Mutation breeding in oil palm : a manual / Fazrin Nur, Brian P.
    Forster, Samuel A. Osei, Samuel Amiteye, Jennifer Ciomas, Soeranto Hoeman,
    Ljupcho Jankuloski.
Description: Boston, MA : CABI, [2018] I Series: Techniques in plantation
    science series ; 4 I Includes bibliographical references and index.
Identifiers: LCCN 2018016830 (print) I LCCN 2018023043 (ebook) I
    ISBN 9781786396235 (ePDF) I ISBN 9781786396228 (ePub) I
    ISBN 9781786396211 (pbk : alk. paper)
Subjects: LCSH: Oil palm--Mutation breeding. I Plant mutation breeding.
Classification: LCC SB299.P3 (ebook) I LCC SB299.P3 N87 2018 (print) I
    DDC 633.8/51--dc23
LC record available at https://lccn.loc.gov/2018016830

ISBN-13: 978178639621 (pbk)

Commissioning editor: Rachael Russell
Editorial assistant: Emma McCann
Production editor: James Bishop

Typeset by SPi, Pondicherry, India
Printed and bound in the UK by Severn, Gloucester.

Series Foreword – Techniques in Plantation Science

Verdant Bioscience, Singapore (VBS), is a new company established in October 2013 with a vision to develop high-yielding, high-quality planting material in oil palm and rubber through the application of sound practices based on scientific innovation in plant breeding. The approach is to fuse traditional breeding strategies with the latest methods in biotechnology. These techniques are integrated with expertise and the application of sustainable aspects of agronomy and crop protection, alongside information and imaging technology which not only find relevance in direct aspects of plantation practice but also in selection within the breeding programme. When high-yielding planting material is allied with efficient plantation practices, it leads to what may be termed 'intensive sustainable' production. At the same time, the quality of new products is refined to give more specialized uses alongside more commodity-based oil production, thus meeting the market demands of the modern world community, but with a minimal harmful footprint. An essential ingredient in all this is having sound and practical protocols and techniques to allow the realization of the strategies that are envisaged.

To achieve its aims, VBS acquired an Indonesian company called PT Timbang Deli Indonesia, with an estate of over 970 ha of land at Timbang Deli, Deli Serdang, North Sumatra, Indonesia, and the group works under the name of 'Verdant'. A central part of this estate, which will be used for important plant nurseries and field trials, is the development of the Verdant Plantation Science Centre (VPSC), to which the operational staff moved in October 2016. A seed production and marketing facility is now established at VPSC for commercial seed sales and the processing of seed from breeding programmes. The centre comprises specialized laboratories in cell biology, genomics, tissue culture, pollen, soil DNA, plant and soil nutrition, bunch and oil, agronomy and crop protection. Field facilities include extensive nurseries, seed gardens and trials (trial sites are also located at various locations across Indonesia). It is the aim of the company to use its existing and rapidly developing intellectual property (IP) to develop superior cultivars that not only have outstanding yield but also are resistant to both biotic and abiotic stresses, while at the same time meeting new market demands. Verdant not only develops and supplies superior planting materials but also supports its customers and growers with a package of services and advice in fertilizer recommendations and crop protection. This is all part of a central mission to promote green, eco-friendly agriculture.

Brian P. Forster and Peter D.S. Caligari
Lead Scientist and Science Strategy Executive Director
Verdant Bioscience

Contents

Acknowledgements

The authors are grateful to all Verdant Bioscience staff, particularly those in the breeding and biotechnology sections, for sharing their knowledge and providing helpful advice in preparing this manual.

Preface

As noted in the Foreword to this series, a central objective in Verdant Bioscience's mission is to develop better cultivars of oil palm, rubber and other plantation crops through plant breeding. Essential to this objective is a wide germplasm base. Conventional breeding is practised by Verdant in the development of new cultivars. However, this can be a time-consuming exercise, especially as the generation time in oil palm is 4–5 years. Verdant is therefore committed to accelerated breeding methods, one of which is mutation breeding. Mutation breeding is not a genetic manipulation (GM) technique and, indeed, it has been practised for many decades; the first mutant cultivar was produced in tobacco in Indonesia in 1934. There are now well over 3000 mutant cultivars in over 200 crop species (http://mvgs.iaea.org) worldwide. It is notable that oil palm is the only oil crop not to have been improved with the aid of mutation breeding, although some experiments have been carried out. In this manual, we give an introduction to mutation breeding and practical protocols on mutation induction and mutant selection in oil palm. Our target audiences are students and researchers in agriculture, plant breeders, growers and end-users interested in the practicalities of producing novel oil palm genotypes for further breeding and commercial exploitation.

Brian P. Forster and Peter D.S. Caligari
Series Editors
February 2018

Introduction

Abstract

A brief and general description of mutation breeding is provided. Mutation has been a successful strategy in producing over 3000 mutant cultivars in over 200 crop species worldwide. Oil palm is one of the few major crop species, and the only oil crop not to have been improved by mutation breeding. However, pioneering work in mutation breeding in oil palm did take place in Ghana in the 1970s. This produced the first M_1 population in oil palm and has only recently been progressed by developing M_2 populations and in discovering mutants for crop improvement.

1.1 Brief History of Plant Mutation Breeding

The history of plant breeding spans centuries. The improvement of crop plants has been continuous since the domestication of species: initially by simple selection of naturally occurring forms; later by deliberate intervention. A landmark for plant breeding was the establishment of the laws of inheritance (Mendel, 1866), which transformed plant breeding from an art into a science. Since then, various plant breeding technologies have had significant impacts on plant breeding. These include: induced polyploidy; interspecific hybridization; chromosome engineering; alien gene introgression; doubled haploidy; F_1 hybrids; transformation; genetic modification (GM); and induced mutation.

The history of plant mutation breeding has been the subject of numerous publications (van Harten, 1998; and, more recently, Forster and Shu, 2012; Kharkwal, 2012; Bado *et al.*, 2015; Bado *et al.*, 2017) and, therefore, only a brief overview is given here. Hugo de Vries is regarded as the founding father of mutation breeding (his life and work is described by Blakeslee, 1935). At the beginning of the 20th century, de Vries published a series of papers on mutation theory (1901, 1903, 1905) in which he advocated its potential in plant breeding. Up to this point, natural spontaneous mutations were the only means of generating new genetic diversity.

Important spontaneous mutants in oil palm include shell-less fruit (Pisifera; Fig. 1.1) and virescent fruit colour (Fig. 1.2). The shell gene (*Sh*) is the most economically important gene in oil palm production and has been the subject of intense genetic studies. The gene has been sequenced and DNA diagnostic markers have been developed and used in determining shell type (Singh *et al.*, 2014). Wild-type plants, known as Dura (*Sh/Sh*), have a thick shell (endocarp) around the kernel of the fruit, whereas mutant types, known as Pisifera (*sh/sh*), have no shell (and often exhibit female sterility). The hybrid between Dura × Pisifera crosses, known as Tenera (*Sh/sh*), has a thin shell and has the most productive fruit in terms of oil content. Tenera types are the commercial planting material (Kelanaputra *et al.*, 2018). The virescent mutant is also of interest, as the change from green to yellow/orange fruits on ripening is more marked than the normal black to orange/red. This can be exploited in selecting fruit bunches at the optimal stage (fully ripe) for oil extraction. Like *Sh*, genetic markers have been developed for the detection of variation in the *Vir* gene. These can be exploited in marker-assisted breeding (Singh *et al.*, 2015).

Fig. 1.1. Fruit types of oil palm: (a) Dura wild-type fruit with a thick shell around the kernel; (b) Pisifera mutant-type, shell-less fruit (the kernel has no shell protection); and (c) Tenera thin-shelled fruit.

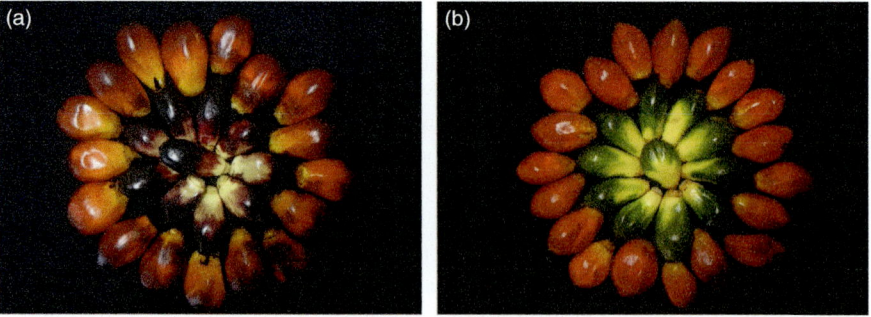

Fig. 1.2. Fruit colour in oil palm: (a) nigrescent wild-type fruit colour change from black unripe fruit (centre) to orange/red ripe fruit (perimeter); (b) virescent mutant-type fruit colour change from green unripe fruit (centre) to yellow/orange ripe fruit (perimeter).

After de Vries, proof of his mutation theory was established, notably by the pioneering work of Lewis John Stadler, who induced mutations in crop plants (barley and maize; Stadler, 1928). A major milestone was the first mutant crop variety, cv. Chlorina in tobacco. This was produced by X-ray irradiation in Indonesia in the 1930s (Tollenaar, 1934). The success in tobacco was followed by the establishment of physical mutation induction techniques (gamma and X-ray) and their worldwide application in a large range of crops, promoted and supported by the Joint FAO/IAEA Division of Nuclear Techniques in Food and Agriculture. Mutation breeding is now integrated into a range of enabling biotechnologies, especially those that aid the detection of desired mutations and further accelerate the plant breeding process.

A history of plant mutation breeding has been given by van Harten (1998) and updated by Kharkwal (2012), who lists five distinct periods:

Period 1 (300 BC to 1894): observation and documentation of early spontaneous mutants.
Period 2 (1895–1920): conceptualization of mutation and mutation breeding.
Period 3 (1926–1959): proof of induced mutations and release of the first commercial mutant varieties.
Period 4 (1964–1990): large-scale application of mutation breeding.
Period 5 (1983 to present): integration of plant mutation with biotechnologies and genomics.

(See Kharkwal (2012) for details of the key events in each of these periods.)

The main factors that contribute to the success of plant mutation breeding are:

- It is fast, much faster than cross-breeding.
- Mutation induction is quick and easy.
- It is cheap: there is no need to invest in specialized equipment, as registered irradiation facilities are available nationally, regionally and internationally (see Chapter 6 of this manual).
- Desired traits can be induced directly into elite lines.
- It is ubiquitous, as it can be applied to any crop and for any trait.
- It is successful: there are well over 3000 registered mutant varieties in over 200 crop species (IAEA, 2017).
- It is NOT a GM technique.

The major limitations in plant mutation breeding are:

- Mutation detection: until recently this has relied entirely on the phenotypic selection of mutant traits and could only be applied to the M_2 generation at the earliest (Shu *et al.*, 2012a). However, new developments in genotypic selection provide more efficient selection systems.

- The length of time needed to develop mutant varieties in perennial crops.
- The space required to grow out large mutant populations; this is particularly problematic for perennial crops such as trees, which take up a lot of space.
- Not all types of useful variation can be induced as mutations.

Much of the early success in mutation breeding was with annual seed-propagated crops, as it is relatively quick to get through the life cycles needed to develop mutant varieties from the initial (M_0) mutant population, usually up to M_{6-8}. Vegetatively propagated crops and perennial crops (including oil palm) have lagged behind, due to the limitations listed above. However, this is changing. New developments in tissue culture offer efficient methods for the selection and development of mutant lines in vegetatively propagated crops (reviewed by Bado *et al.*, 2017). Also, genomic screening in early generations provides a means of short-cutting generations and accelerating mutation breeding in perennial crops (see Chapter 4 of this manual).

1.2   General Principles of Plant Mutation Breeding

Mutation breeding typically aims to improve elite lines or superior cultivars for desired traits, such as resistance to diseases and pests, tolerance to biotic and abiotic stresses, quality, yield and agronomic traits (height, flowering time, etc.), and in the case of oil palm, novel traits for mechanical harvesting. Mutation induction may be brought about by physical, chemical or biological mutagens. Physical mutagenesis, especially gamma irradiation, is a proven and widely used method in mutation breeding. Plant materials are exposed to irradiation for various times at a dose that provides the optimal mutation frequency; this is provided by a radio-sensitivity test (see Chapter 3 of this manual). Irradiation is usually performed in a contained gamma cell: this is generally used for small plant materials, for example seeds and *in vitro* cuttings. Gamma greenhouses and gamma fields are available for whole-plant irradiation (Shu *et al.*, 2012a; Bado *et al.*, 2015). A detailed account of the physics, chemistry and biology involved in transmitted energy, DNA damage and DNA repair in mutation induction is given by Lagoda (2012).

The impact of induced mutation using gamma irradiation has been significant, producing over 60% of mutant cultivars among the over 3000 officially released mutant cultivars (Bado *et al.*, 2015; IAEA Mutant Variety Database: http://mvgs.iaea.org). Mutagenesis is applicable to all plant species and has remained popular for close to a century as a result of its effectiveness, simplicity, usability on a small or large scale and its economic impact (Saddiqui and Khan, 1999; Ahloowalia *et al.*, 2004; Shu *et al.*, 2012b; Bado *et al.*, 2015).

Mutation breeding offers good prospects in crop improvement, as witnessed by over 3000 mutant cultivars that have been officially released from over 200 crop species (IAEA Mutant Variety Database: http://mvgs.iaea.org).

Mutation breeding is classed as a conventional method; it is not a GM method and is without the regulatory restrictions faced by genetically modified materials (Maluszynski *et al.*, 2000). However, oil palm is not found in the list of mutant cultivars released and remains a major crop, and the only oil crop that has not been improved by mutation breeding. Some of the challenges of oil palm mutation breeding include its perennial nature, provision of good growing conditions for M_1 populations (shade, irrigation, space and time), and its heterozygosity (as a result of its outbreeding reproductive system). Another issue with oil palm is that it has one meristem, it does not branch; therefore, the tricks that are commonly used in tree crops to overcome the juvenile stage, such as irradiating buds from mature branches, are not possible. The generation time factor (4–5 years from seed to seed) remains a major issue to be overcome in mutation breeding, as normally mutation detection is performed in the M_2 generation at the earliest. M_1 phenotypic screening is limited to dominant traits (with height and virescent fruit as examples) or dominant allele knockouts at heterozygous loci and genotypic screens (detecting sequence changes). Mutant screening is not carried out until the M_2 generation at the earliest, as M_1 plants are often weak and suffer from chimeras and physiological perturbations caused by the mutagenic treatment. However, although phenotypic screening is not recommended in the M_1, it is possible to screen genotypically in the M_1, and this saves many years in rearing an M_2 population and selecting plants for further development, as well as saving valuable nursery and field space.

1.3 Breeding Limitations in Oil Palm

Oil palm (*Elaeis guineensis* Jacq.) is highly productive compared to other oil crops, and it is capable of meeting the growing world demand for vegetable oils, projected to reach 240 million tonnes (Mt) by 2050 (Corley, 2009). Current production is based on increased acreage and there is now a need to produce more sustainable yields, particularly by higher yields per land area. Plant breeding requires the genetic variation of useful traits for crop improvement. Often, however, the desired variation is lacking. Mutagenic agents, such as radiation and certain chemicals, can be used to induce mutations and generate genetic variation from which the desired mutants may be selected. Mutation induction has become a proven method of creating variation within a crop. It offers the possibility of inducing desired attributes that either cannot be found in nature or have been lost during crop evolution. Breeding for improved plant cultivars is based on two principles: genetic variation and selection. In mutation breeding, this can be altered to: mutation induction and mutant selection. Induced mutagenesis has been practised with great success in crop breeding programmes throughout the world since the 1930s (Ahloowalia *et al.*, 2004; Bado *et al.*, 2015).

1.4   Mutation Breeding in Oil Palm

Oil palm is a diploid: this is an advantage, as induced mutants will not be masked by homologous or homoeologous genes in another genome, a problem encountered in polyploid crops. Other advantages include the large numbers of seed produced from one female flower bunch (300–3000); in addition, the pollen is both abundant and storable and can be used as a target for mutation induction. Recently, haploid production methods have been developed for oil palm (Nelson *et al.*, 2009; Dunwell *et al.*, 2010), and these now offer a new target for mutation induction, as homozygous mutants can be produced instantly by the conversion of haploids to doubled haploids. This is significant, as the vast majority of induced mutations are recessive and therefore their phenotype cannot be seen until the mutant alleles are made homozygous.

In 1977, the Oil Palm Research Institute (OPRI) of the Council for Scientific and Industrial Research (CSIR) in Ghana initiated work with the Ghana Atomic Energy Commission (GAEC) to use gamma rays to promote the germination of oil palm seeds and pollen. The work was spearheaded by Dr J.B. Wonkyi-Appiah (former Director of OPRI). Among all the electromagnetic radiations, gamma rays (which are ionizing radiation) have proven to be the most effective, having the energy level of about 10 kilo electron volts (keV) to several hundred keV. Gamma rays have more penetrating ability than alpha and beta rays. In crop improvement, genetic variability has occurred in irradiated seeds, and plant breeders have been able to select new genotypes with improved characteristics such as salinity tolerance, disease resistance, grain yield and quality (Ashraf *et al.*, 2003; Shu *et al.*, 2012b; Bado *et al.*, 2015).

The gamma irradiation treatments, 10–50 Gy for both pollen and seeds, used by OPRI and GAEC in 1977 were comparatively low for mutation breeding purposes. However, this may not be a disadvantage, as semi-dwarfism and other desirable mutant traits often occur at relatively high frequencies and may appear in subsequent generations (Forster *et al.*, 2012). As a consequence of this work, OPRI has produced the first induced mutant populations (M_1) in oil palm and has a population of 575 M_1 palms obtained from irradiated seeds and irradiated pollen. The M_1 population represents a unique and very valuable resource for oil palm improvement. As stated above, M_1 plants usually suffer from chimeras and physiological disorders, and therefore cannot be used for phenotypic screening of heritable traits. The chimeras are dissolved through meiosis, and therefore there is the need to produce the M_2 population and subsequent mutant generations. In addition, if the traits appear in the M_2 and advanced generations, they generally must be heritable (Prina *et al.*, 2012; Ukai and Nakagawa, 2012). The oil quality of all oil crops has been improved by mutation breeding, with the notable exception of oil palm (Vollmann and Rajcan, 2010).

Oil palm takes 4–5 years to flower from planting a seed; that is, at least 4–5 years to produce the M_2 seed from sowing the M_1. The Ghana M_1 population represented the only known induced mutant population in oil palm until Verdant began a mutation breeding programme in 2017. The OPRI M_1 trees are currently about 40 years old, and in general are phenotypically normal (Fig. 1.3) and based on the regular production of fruit bunches. Good trees were selected from various irradiation regimes for pollen collection for subsequent self-pollination to produce M_2 seeds; this was done in 2010. The seeds were processed and germinated and transferred to the nursery stage in 2017 (Fig. 1.4). Pollen collection and selfing started in 2010 and was supported by the International Atomic Energy Agency (IAEA) and Sumatra Bioscience, Indonesia. The IAEA continues to support this programme through capacity building, training and equipment (IAEA Technical Cooperation Project GHA5036). The M_2 seedlings are currently being screened in the nursery for resistance to fusarium wilt disease, semi-dwarfism, drought tolerance, vegetative characters and any abnormalities. Screening will not only be phenotypic, as molecular methods will also be employed to screen for traits genotypically; traits such as shell thickness, virescent fruit, oil quality, mantled fruit and other traits of interest where gene sequence data are known.

Plant research has been revolutionized as a result of the development of biotechnology tools, particularly in the area of molecular breeding. Molecular breeding is a concept in which conventional breeding is assisted by DNA markers (Rafalski and Tingey, 1993). The genome of oil palm has been sequenced (Singh *et al.*, 2013) and it is now practicable to employ

Fig. 1.3. Mature M_1 populations at OPRI, Ghana.

Fig. 1.4. M$_2$ seedlings in large polybags in a nursery at OPRI, Ghana.

sequence data to screen for mutations in target genomic regions in mutant populations. The main advantage of marker-assisted selection is that plants may be selected early in plant development (at M$_1$), before the trait of interest is expressed (shell type, fruit colour, oil quality, mantled fruits, etc.), and this in turn can reduce greatly the time required to bring new cultivars to the market (Mazur and Tingey, 1995).

Harvesting oil palm is expensive in terms of manual labour: it is an arduous task compared to the ease of combine-harvesting arable crops, which has been assisted greatly by the introduction of semi-dwarf mutant cultivars (Corley and Tinker, 2003; Bado *et al.*, 2015). Radical changes to the oil palm tree architecture are needed to enable the development of harvesting machines, and this involves novel traits; for example, palm height, late fruit abscission, long bunch stalk and fruit colour marker for ripeness (Le Guen *et al.*, 1990). M$_2$ oil palms will be grown in the field and at maturity screened for other traits of value, such as those required for mechanical harvesting.

References

Ahloowalia, B.S., Maluszynski, M. and Nichterlein, K. (2004) Global impact of mutation derived varieties. *Euphytica* 135, 187–204.

Ashraf, M., Cheema, A.A., Rashid, M. and Qamar, Z. (2003) Effect of gamma rays on M$_1$ generation in Basmati rice. *Pakistan Journal of Botany* 35, 791–795.

Bado, S., Forster, B.P., Nielen, S., Ali, A.M., Lagoda, P.J.L. *et al.* (2015) *Plant Mutation Breeding: Current Progress and Future Assessment. Plant Breeding*

Reviews 39. (Janick, J. (ed.)). Wiley-Blackwell, Hoboken, New Jersey,Chapter 2, pp. 23–88.

Bado, S., Yamba, N.G.G., Sesay, J.V., Laimer, M. and Forster, B.P. (2017) Plant mutation breeding for the improvement of vegetatively propagated crops: successes and challenges. *CAB Reviews* 12, No 028 (online ISSN 1749-8848).

Blakeslee, A.F. (1935) Hugo de Vries 1848–1935. *Science* 81, 581–582.

Corley, R.H.V. (2009) How much palm oil do we need? *Environmental Science & Policy* 12, 134–139, DOI: 10.1016/j.envsci.2008.10.011.

Corley, R.H.V. and Tinker, P.B. (2003) *The Oil Palm*, 4th edn. John Wiley & Sons, Oxford, DOI: 10.1002/9780470750971.

de Vries, H. (1901) *Die Mutationstheorie I*. Veit & Co, Leipzig, Germany.

de Vries, H. (1903) *Die Mutationstheorie II*. Veit & Co, Leipzig, Germany.

de Vries, H. (1905) *Species and Varieties: Their Origin by Mutation*. The Open Court Publishing Company, Chicago, Illinois.

Dunwell, J.M., Wilkinson, M.K., Nelson, S.P.C., Wening, S., Sitorus, A.C. *et al.* (2010) Production of haploids and doubled haploids in oil palm. *BMC Plant Biology* 10, 218–243. Available at: http://www.biomedcentral.com/1471-2229/10/218 (accessed 14 March 2018).

Forster, B.P. and Shu, Q.Y. (2012) Plant mutagenesis in crop improvement: basic terms and applications. In: Shu, Q.Y., Nakagawa, H. and Forster, B.P. (eds) *Plant Mutation Breeding and Biotechnology*. CAB International and FAO, Wallingford, UK, Chapter 1, pp. 9–20.

Forster, B.P., Franckowiak, J.D., Lundqvist, U., Thomas, W.T.B., Leader, D. *et al.* (2012) Mutant phenotyping and pre-breeding in barley. In: Shu, Q.Y., Nakagawa, H. and Forster, B.P. (eds) *Plant Mutation Breeding and Biotechnology*. CAB International and FAO, Wallingford, UK, Chapter 25, pp. 327–346. ISBN 19: 978-178064-085-3 (CABI); ISBN 13:978-925107-022-2 (FAO).

IAEA (2017) IAEA Mutant Variety Database. Available at: http://mvgs.iaea.org (accessed 14 March 2018).

Kelanaputra, E.S., Nelson, S.P.C., Setiawati, U., Sitepu, B., Nur, F. *et al.* (2018) *Oil Palm Seed Production: A Manual*. CAB International, Wallingford, UK.

Kharkwal, M.C. (2012) Brief History of Plant Mutagenesis. In: Shu, Q.Y., Nakagawa, H. and Forster, B.P. (eds) *Plant Mutation Breeding and Biotechnology*. CAB International and FAO, Wallingford, UK, Chapter 2, pp. 21–30.

Lagoda, P.J.L. (2012) Effects of radiation on living cells and plants. In: Shu, Q.Y., Nakagawa, H. and Forster, B.P. (eds) *Plant Mutation Breeding and Biotechnology*. CAB International and FAO, Wallingford, UK, Chapter 11, pp. 123–134.

Le Guen, V., Ouattara, S. and Jacquemard, J.C. (1990) Oil palm selection with a view to easier harvesting. Initial results. *Oléagineux* 45, 523–531.

Maluszynski, M.K., Nichterlein, K., van Zanten, L. and Ahloowalia, B.S. (2000) Officially released mutant varieties – the FAO/IAEA Database. *Mutation Breeding Review* 12, 1–84.

Mazur, B.J. and Tingey, S.V. (1995) Genetic mapping and introgression of genes of agronomic importance. *Current Opinion in Biotechnology* 6, 175–182.

Mendel, G. (1866) Versuche über Pflanzenhybriden. *Verhandlungen des naturforschenden Vereines in Brünn*, Bd. IV für das Jahr 1865, Abhandlungen, 3–47.

Nelson, S.P.C., Wilkinson, M.J., Dunwell, J.M., Forster, B.P., Wening, S. *et al.* (2009) Breeding for high productivity lines *via* haploid technology. In: *Proceedings of Agriculture, Biotechnology and Sustainability Conference*, PIPOC 9–12 November 2009, Kuala Lumpur, Malaysia, Volume 1, pp. 203–225.

Prina, A.R., Landau, A.M. and Pacheco, M.G. (2012) Chimeras and mutant gene transmission. In: Shu, Q.Y., Forster, B.P. and Nakagawa, H. (eds) *Plant Mutation Breeding and Biotechnology*. CAB International and FAO, Wallingford, UK, pp. 181–189.

Rafalski, J.A. and Tingey, S.V. (1993) Genetic diagnostics in plant breeding: RAPDs, microsatellites and machines. *Trends in Genetics* 9, 275–280.

Saddiqui, B.A. and Khan, S. (1999) *Breeding in Crop Plants: Mutations and in vitro Mutation Breeding*. Kalyani Publishers, Ludhiana, India, pp. 35–56.

Shu, Q.Y., Forster, B.P. and Nakagawa, H. (2012a) Principles and applications of plant mutation breeding. In: Shu, Q.Y., Nakagawa, H. and Forster, B.P. (eds) *Plant Mutation Breeding and Biotechnology*. CAB International and FAO, Wallingford, UK, Chapter 24, pp. 301–326.

Shu, Q.Y., Forster, B.P. and Nakagawa, H. (2012b) *Plant Mutation Breeding and Biotechnology*. CAB International and FAO, Wallingford, UK, 608 pp. ISBN-19: 978-178064-085-3 (CAB International); ISBN-13:978-925107-022-2 (FAO).

Singh, R., Ong-Abdullah, M., Low, E.T.L., Manaf, M.A.A., Rosli, R. *et al.* (2013) Oil palm genome sequence reveals divergence of interfertile species in old and new worlds. *Nature* 500, 335–341.

Singh, R., Ti, L.L.E., Ooi, C.L.L., Ong-Abdullah, M., Manaf, M.A.A. *et al.* (2014) *SureSawit*™ *Shell – A Diagnostic Assay to Predict Oil Palm Fruit Forms*. MPOB (Malaysian Palm Oil Board), TT No 548 (June), 656.

Singh, R., Ooi, L.C.C., Low, L.E.T., Ong-Abdullah, M., Nagappan, J. *et al.* (2015) *SureSawit* ™ *Vir – A Diagnostic Assay to Predict Colour of Oil Palm Fruits*. MPOB (Malaysian Palm Oil Board), TT No 568 (June), 697.

Stadler, L.J. (1928) Genetic effects of X-rays in maize. *Proceedings of the National Academy of Sciences of the United States of America* 14, 69–75.

Tollenaar, D. (1934) Untersuchungen über Mutations bei Tabak. *Genetica* 16, 111–152.

Ukai, Y. and Nakagawa, H. (2012) Strategies and approaches in mutant population development for mutant selection in seed propagated crops. In: Shu, Q.Y., Forster, B.P. and Nakagawa, H. (eds) *Plant Mutation Breeding and Biotechnology*. CAB International and FAO, Wallingford, UK, pp. 209–221.

van Harten, A.M. (1998) *Mutation Breeding: Theory and Practical Applications*. Cambridge University Press, Cambridge.

Vollmann, J. and Rajcan, I. (eds) (2010) *Oil Crops. Handbook of Plant Breeding*. Springer Verlag, New York.

Health and Safety Considerations and Guidelines

2

Abstract

Mutation induction is the most hazardous part of mutation breeding. The application of physical or chemical mutagens as well as the general handling of mutant materials in mutation breeding has associated high health and safety risks. Poor practice can be hazardous, resulting in minor or serious personal injuries, and even fatal accidents. Therefore, the use of standard health and safety protocols as guides in mutation induction activities is an absolute necessity. It is important to abide by the rules and regulations of mutation induction in order to avoid, or reduce to the barest minimum, injuries and accidents, as well as associated financial costs. With respect to chemical mutagens, these need specialist waste disposal systems after use. **It is essential that mutation induction is carried out by specially trained personnel, and it is recommended that this is done in specialized institutes, or at least laboratories.** International, regional and national service facilities are available (Chapter 6 of this manual). Certain rules and regulations constituting the operation guidelines may be generally applicable; however, depending on local requirements, some rules and regulations may be specific to a particular region or country. In certain jurisdictions, negligence in the strict adherence to health and safety protocols may incur sanctions such as fines, jail terms, or an embargo on field and laboratory research or commercial operations. This chapter highlights the standard operating protocols on health and safety issues relating to mutation induction for breeding in oil palm. Once mutation treatment has been carried out, subsequent steps (mutation detection, mutant line development and breeding) are relatively safe, but adherence to good laboratory, glasshouse and field practices is essential.

2.1 Health and Safety in Mutation Induction and Radio-sensitivity Testing

Mutagens are used for two key processes in plant mutation breeding: (i) radio-sensitivity testing to determine the optimal dose treatment; and

(ii) mutagenesis to produce a mutant population. Both procedures should be performed in specialized laboratories by trained staff. Seed is normally the choice of material for mutation treatments, but in oil palm the thick shell around the kernel is a barrier to X-rays and, to a lesser extent, gamma rays; also, oil palm seed suffers from dormancy. For this reason, the preferred target material is germinated seed (M_0). Oil palm has an advantage in that germinated seeds are readily available from seed producers. Alternatively, seed may be germinated in-house (Kelanaputra *et al.*, 2018). Radio-sensitivity testing is required to determine the optimal irradiation treatment; this involves exposure to irradiation and should only be performed in specific laboratories. After treatment, the M_1 seed of oil palm (for radio-sensitivity testing and mutation detection) are grown up in benign conditions in a nursery. The mutant population is screened for desired traits at various stages in plant development (nursery and field); mutant lines are selected and advanced with the aim of producing new breeding lines and cultivars. Each stage in mutation induction, mutation detection and mutant line development has its own set of health and safety considerations. It is therefore necessary that the relevant health and safety measures or precautions relating to nursery and field operations, as well as laboratory activities during material sampling, mutagen treatment or mutant plant establishment, are observed and adhered to.

2.2   Health and Safety in the Field

Harvesting or sampling experimental materials in oil palm may include the inflorescence (male and female), fruit bunches or leaves. Harvesting is an arduous, labour-intensive activity that presents safety risks to workers (Bamidele, 2015). A major safety threat to the personnel involved in fruit sampling activities is the height and stability of old palm trees. Trees 20 years or older can reach heights of over 15 m. The height of palms is an important consideration because inflorescences and bunches are located in the canopy of the trees. During manual harvesting of palm fruit bunches, a worker may fall from the palm tree, leading to broken limbs or even paralysis. A similar fate could befall assistant workers on the ground in accidents where cut bunches fall down. It is therefore safer, and minimal health risks are encountered, when sampling of inflorescences, bunches or leaves is done from young and/or short palms. This is recommended whenever possible. However, in cases where the only option available is to sample materials from older and taller palm trees, then a number of health and safety procedures must be deployed. It is important to note that injury may also be sustained in the form of particles entering the eyes. The spikelet of the palm fruit bunch may pierce the worker's eyes. Stings and bites from nuisances such as bees and snakes are also very important considerations.

Field precautions and safety measures

- *Accessing tall palms safely.* Workers must dress in long-sleeved shirts, trousers or overalls and have gloves on their hands to reduce skin exposure. Oil palm is a tropical crop and therefore grows where sunlight is intense, and so the use of sunblock and shade hats are important considerations. It is also important that appropriate boots or footwear are used. Protective clothing minimizes stings and bites from insects, snakes and other nuisance pests. Wearing a helmet is very important as a protection against falling objects such as fruit bunches and palm fronds (Bamidele, 2015). Harnesses must be checked to ensure that they are in good condition and must be worn by workers to minimize the chances of falling from the tall trees. Ladders must be used to ease tree climbing and must be secured to the tree. Ladders must be checked periodically for integrity. It is advisable for at least two workers to assist each other when engaging in such field activities.
- *Chemical applications.* In order to safeguard uncontrolled pollination to enable the production of seeds from controlled crosses and of specific genetic background, hazardous chemicals such as formalin, insecticides and fungicides are sprayed on to the isolated female inflorescence (Setiawati *et al.*, 2018). Furthermore, it may be important to quicken the loosening of fruits from the spikelets of freshly harvested bunches to within 24 h, to enable seed sampling and preparation. In such instances, bunches are sprayed with ethephon solution to induce the loosening of fruits from spikelets. In undertaking these practices, gloves, respirators and other protective clothing must be worn. This is required to prevent skin contact and chemical inhalation. Any contact with hazardous chemicals should be followed immediately by washing and first aid.
- *Appropriate working tools.* Appropriate tools, including sharp knives, cutlasses, chisels or sickles, should be used and with the correct techniques. For additional safety, tools must be transported and carried in appropriately approved containers. Ensure that the tools are cleaned after use and stored in a designated area or store room.
- *Emergency procedures.* First aid training and access to a first aid box, telephone numbers of emergency services and compliance with in-house emergency procedures are essential.

2.3   Health and Safety for Irradiation Treatments

Physical agents are used to induce heritable mutations artificially in plants. The most widely used physical agent is ionizing radiation, especially gamma rays, to treat seeds and other plant materials (Harun, 2001; Bado *et al.*, 2015). If radiation is not used properly under controlled conditions, it can be extremely hazardous to the health of workers. It is mandatory for the irradiation facilities that provide a service for the treatment of seeds and other materials with gamma radiation to have and to adhere to standard operating protocols to safeguard the health and safety of workers and other people around. Samples for irradiation treatment are normally handed over to specialized staff, and the mutation breeder is normally not directly involved. Specialized irradiation facilities are normally available nationally, regionally or internationally (Spencer Lopez *et al.*, 2018), and plant materials may be sent to these using standard material transfer agreements and compliance with quarantine regulations; for example, the provision of phytosanitary certificates. Thus, health and safety issues, pre-irradiation and post-irradiation, are the primary concern of the mutation breeder.

The IAEA provides recommendations for protecting people and the environment in a continually updated series on basic safety standards (see, for example, Boal *et al.*, 2013). They are based on knowledge of radiation effects and on established principles of radiation protection recommended by the International Commission on Radiological Protection (Adlienė and Adlytė, 2017).

2.4   Health and Safety in the Laboratory

The induction of mutation in oil palm requires various laboratory procedures and the use of chemicals for seed preparation. In oil palm seed, preparation mainly involves the removal of the mesocarp and germination of the seeds under established seed laboratory conditions. The seed is composed of a kernel containing the endosperm and embryo, which is protected by a thick shell (endocarp). The thick shell is a barrier to mutation treatments and the oil palm seed suffers from dormancy; thus, the preferred material for mutation induction is germinated seed. Germinated seed is available commercially or may be produced oneself (Kelanaputra *et al.*, 2018). It is therefore important to note the necessary laboratory health and safety requirements (Barker, 2005), as well as vital precautions for the safe handling of the chemicals used. This is achieved by adherence to good laboratory and chemical handling practices.

Laboratory precautions and safety measures

- *Protection in the laboratory.* A laboratory coat should be put on before entry into the laboratory and should be removed when leaving. Full foot-covering, low-heeled shoes should be worn. This provides protection to the individual from the samples being worked on, as well as protection from contamination with laboratory materials for other people inside or outside the laboratory.
- *Accessory protective clothing.* Appropriate protective items such as glasses or goggles, face and nose shields and gloves should be worn,

especially when using any hazardous agents, as well as when using equipment. These items prevent direct skin contact and the inhalation of chemicals. Direct skin contact or inhalation of chemicals used for laboratory techniques such as DNA and RNA analysis, flow cytometry or tissue culture must be avoided. Ear plugs may be necessary when using high-sounding equipment such as sonicators for tissue disruption in DNA isolation. Laboratories should be fitted with a first aid box, eyewash and a shower for emergency washing. Remove laboratory clothing before returning to communal areas (the office, for example).

- *Work hygiene.* Keep the work area clear of all materials except those needed for the work being undertaken. Materials should be kept away from equipment that requires air flow or ventilation to prevent over-heating–in laminar flow hoods, for instance.
- *Laboratory do nots.* Never pipette anything by mouth: a bulb must be used. Solvents should never be 'smelled'. Labels on solvent bottles must be read carefully and understood to identify the contents and the hazards. Never taste anything.
- *Safe waste disposal.* Personnel should be conversant with the laboratory's standard operating procedures (SOPs) for waste disposal. Individuals are responsible for the proper disposal of used material, if any, in appropriate containers. Spills must be cleaned up immediately using the agreed procedures. Waste such as gels, staining dyes, scalpel blades or broken glassware must be disposed of in proper containers. Contaminated tissue cultures must be decontaminated by autoclaving before disposal. Check glassware for cracks and chips each time before and after use. Cracks can cause glassware to fail during use, which can cause serious injury.
- *Emergency procedures.* Personnel should be aware of emergency procedures: firefighting, emergency exits and emergency telephone numbers, and the location of assembly areas, fire extinguishers and first aid/first aiders and medical facilities. Maintain clear access to all exits, fire extinguishers, electrical panels, emergency showers and eyewashes.

2.5 Health and Safety in Handling Hazardous Chemicals

Compared to irradiation, mutagenic chemicals are also used for mutation induction. Chemical mutagens have been found to induce less toxicity but effect higher frequency of mutations in plants. Mutagenic chemicals have, therefore, been useful as research tools for the induction of different types of mutations (Harun, 2001; Ingelbrecht *et al.*, 2018). Conducting chemical mutagenesis exposes the worker(s) involved to the particular chemical. The most widely used chemical mutagen is ethyl methylsulfonate (EMS). Other chemical mutagens include acridine, proflavines, basic analogues, diethyl sulfur (DES), ethylamine (EA) and nitrous methylurethane (NMUT). Careful management of these chemicals will protect the people using them and the general staff, as well as other personnel.

How to control hazardous chemicals and other materials

- *Use of alternatives*. In order to reduce the risks of hazardous chemicals or materials, it is important to consider using less hazardous alternatives where possible, for example, physical mutagens. However, in case that is not possible, it is important to apply control measures.
- *Control measures*. Treat every chemical as if it were hazardous. Make sure all chemicals are labelled clearly and correctly with the substance name, concentration, date and name of the individual responsible.
- *Use of fume chamber*. Volatile and flammable compounds must be used only in a fume hood or chamber. Procedures that produce aerosols should also be performed in a fume hood, to prevent the inhalation of hazardous material. Suitable personal protective clothing and equipment, including respirators, must be worn.
- *Posting of hazard areas*. Areas containing lasers, biohazards, radioisotopes and carcinogens should be signposted accordingly. Labels should be removed when the hazards are no longer present.
- *Material safety data information*. Be aware of hazards relating to the chemicals used in the laboratory and read their Material Safety Data Sheet (MSDA). This provides information on health and safety, first aid, fire and explosion risks, disposal, how to clean up spillages, and handling and storage of a particular chemical.
- *Chemical waste management*. Solid waste must be separated from liquid chemical waste, which in itself may need different storage for various types of chemicals – do not mix unless it is clearly safe to do so. Put lids on both solid and chemical waste storage bins. Transfer points must be enclosed and conveying systems established. Never return chemicals to reagent bottles. Wash exposed skin in the case of skin contact during waste disposal and apply moisturizing cream after drying.
- *Personnel medical check*. The health of workers must be monitored closely to detect early symptoms and seek appropriate medical advice. Those working with irradiation are required to wear radiation monitoring badges, which need regular inspection. All such workers should have a yearly medical examination specifically with this risk in mind.

References

Adlienė, D. and Adlytė, R. (2017) *Dosimetry Principles, Dose Measurements and Radiation Protection*. Institute of Nuclear Chemistry and Technology, Warsaw, 55 pp.

Bado, S., Forster, B.P., Nielen, S., Ali, A.M., Lagoda, P.J.L. *et al.* (2015) *Plant Mutation Breeding: Current Progress and Future Assessment*. *Plant Breeding Reviews* 39. (Janick, J. (ed.)). Wiley-Blackwell, Hoboken, New Jersey, Chapter 2, pp. 23–88.

Bamidele, J. (2015) Occupational hazards and their effect on the health and socio-economic status of local palm oil processors in Delta State, Nigeria. *Annals of Agricultural and Environmental Medicine* 22(3), 483–487.

Barker, K. (2005) *At the Bench: A Laboratory Navigator*. Cold Spring Harbor Press, New York.

Boal, T., Colgan, P.A. and Czarwinski, R. (2013) International Basic Safety Standards – protecting people and the environment. *Radioprotection* 48, S27–S33.

Harun, A.R. (2001) The effective use of physical and chemical mutagen in the induction of mutation for crop improvement in Malaysia. *JAERI-Conference 2001-003*, Hanoi, Vietnam, 9–13 October 2000. Japan Atomic Energy Research Institute, pp. 111–122.

Ingelbrecht, I., Jankowicz-Cieslak, J., Szurman, M., Till, B.J. and Szarejko, I. (2018) Chemical mutagenesis. In: Spencer Lopez, M., Forster, B.P. and Jankuloski, L. (eds) *Manual of Plant Mutation Breeding*. FAO/IAEA publication. FAO, Rome, Chapter 2 (in press).

Kelanaputra, E.S., Nelson, S.P.C., Setiawati, U., Sitepu, B., Nur, F. *et al.* (2018) *Seed Production in Oil Palm: A Manual*. CAB International, Wallingford, UK.

Setiawati, U., Sitepu, B., Nur, F., Forster, B.P. and Dery, S. (2018) *Crossing in Oil Palm: A Manual*. CAB International, Wallingford, UK.

Spencer Lopez, M., Forster, B.P. and Jankuloski, L. (2018) *Manual of Plant Mutation Breeding*. FAO/IAEA publication, FAO, Rome (in press).

Radio-sensitivity Testing

3

Abstract

Before physical mutagenesis can be applied for plant breeding purposes, the optimum irradiation dose needs to be determined. This is done using a radio-sensitivity test in which a range of dose treatments are applied to the target material. After irradiation treatments, the materials are grown out and recordings made for growth reduction (RD) and survival (LD). Normally, values of RD_{30-60} or LD_{30-60} are selected. Here, the target materials for radio-sensitivity testing are germinated seed.

3.1   Need for Radio-sensitivity Testing

Radio-sensitivity, or determination of the optimum dose of radiation, is a term describing a relative measure of the quantity of recognizable effects of a radiation exposure on the irradiated material (Owoseni *et al.*, 2007). The plant breeder considers dose optimization as the first step in inducing a mutation breeding programme (Rohani *et al.*, 2012). The optimal dose is one that produces sufficient mutation events so that there is a reasonable chance of the desired mutation occurring in the mutant population produced, but with a low background mutational load. Radio-sensitivity tests are performed with a wide range of doses to estimate the dose that produces effects normally between LD_{30} and LD_{50}, and RD_{30} and RD_{50}, when using a lethal parameter for the progeny and biomass parameter, respectively. These ranges have been observed to preserve the species integrity of the M_1 (first mutant population) with the least possible unintended damage, but may be altered depending on a range of factors–for example, the breeding system of the species (Mba *et al.*, 2010; Kodym *et al.*, 2012).

3.2 Choice of Material for Radio-sensitivity Testing

Limited data are available on the response of oil palm to mutagenic treatments, including gamma irradiation. Therefore, an appropriate dose of radiation needs to be established on target plant materials before large-scale mutagenesis is undertaken (Tshilenge-Lukanda *et al.*, 2012). Oil palm is propagated by seed, and seed is a useful and easy material to work with for mutation induction (Ghanim *et al.*, 2018). Oil palm seed, however, suffers from dormancy, and it is preferable to work with uniform batches of germinated seed (see Kelanaputra *et al.*, 2018, for methods in seed germination). Stages in oil palm germination are given in Table 3.1

3.3 Choice of Irradiation Dose for Mutation Induction

The choice of irradiation dose is critical for the success of mutation induction. For instance, while a high dose may give a high mutation induction frequency, it is accompanied by many undesirable mutations throughout the genome. Additionally, mutation induction aims at optimizing genetic variation with minimal plant injuries, meaning that a balance has to be found between achieving the desired mutations and maintaining the integrity of the majority of the genome constitution of mutated material (Ndofunsu *et al.*, 2015) and the growth and vigour of the exposed material. The frequency and density of induced mutation depend on the radiation type, the applied dose and the material targeted. The plant species, the ploidy level, differences in developmental stage, physiological condition, etc., may be the cause of differences in the response to radiation, and therefore standardization of the target materials is crucial (van Harten, 1998). Radio-sensitivity testing has been performed on germinated oil palm seeds, as shown in Fig. 3.1.

Fig. 3.1. Seedling development 12 weeks after gamma-ray irradiation of germinated seed of oil palm with increasing dose treatments, from left to right: 0 (control), 5, 10, 15, 20, 30 and 100 Gy.

Table 3.1. Stages of oil palm seed germination with factors that may influence radio-sensitivity.

Germination	Description	Picture
Stage 1	Oil palm seed is a nut with a dense shell, which can vary in thickness within and between genotypes. Variation in shell thickness, water content and oil content of the kernel are factors that influence radio-sensitivity.	
Stage 2	Synchronous germination after various seed treatments (Kelanaputra *et al.*, 2018). At this stage, shoots begin to emerge but are not fully exposed. This is an intermediate stage between seed and germinated seed.	
Stage 3	Both the roots and shoots begin to emerge, with lengths at around 1 cm, but the meristem is not fully exposed.	
Stage 4	Emerged roots and shoots with length above 1 cm; the apical meristem is fully exposed and ready for gamma irradiation treatment. At this stage, it is safe to transport seed with minimal damage. Seeds with shoots and roots longer than 2 cm are susceptible to damage.	

Considering a wide range of radiation doses, the radiation dose is directly proportional to the frequency of mutation induced by radiation. The expression of morphological changes is observed in the sample with a higher dosage of radiation. Very low levels of irradiation often produce positive effects such as increased germination rates and increased seedling vigour, while high levels of irradiation cause severe damage, lack of growth and even death of the seedling.

Mutagenic treatments disrupt both the physiology (reduce growth pro-
moters, increase growth inhibitors and metabolic status) as well as the genetic
constitution. High irradiation doses can be lethal, thus resulting in few plants for
selection. This in turn limits the success of artificial selection in the subsequent
mutant generations to identify useful mutants. Conversely, low irradiation treatments are accompanied by early emergence, increased per cent of germination
and field survival with healthy and vigorous seedlings. However, this is mostly
physiological and is associated with a low mutation frequency and with reduced
chances of being able to select successfully for a target mutant trait.

3.4   RD and LD Determinations for Oil Palm Seed

The best germination stage is Stage 3 (see Table 3.1), as at this stage the germinating seed can be transported from the breeding site to the irradiation facility
and back again with minimal damage. Shoots and roots of germinating seed
larger than 1 cm are prone to damage in transport; therefore, the transfer to and
from the irradiation facility should be done as quickly as possible, preferably
the same day, but within a week, as germinating seeds grow quickly. According
to Mba *et al.* (2010), the dose of mutagen that is regarded as optimal is the one
that achieves the optimum mutation frequency; this is usually a lethal dose of
30–50%. Optimum dose treatments for germinated oil palm seed were established by fitting a curve to the data on seedling survival after gamma treatment
(Fig. 3.2). From Fig. 3.2, the optimum dose treatment (LD_{30-50}) is 11 and 14 Gy.

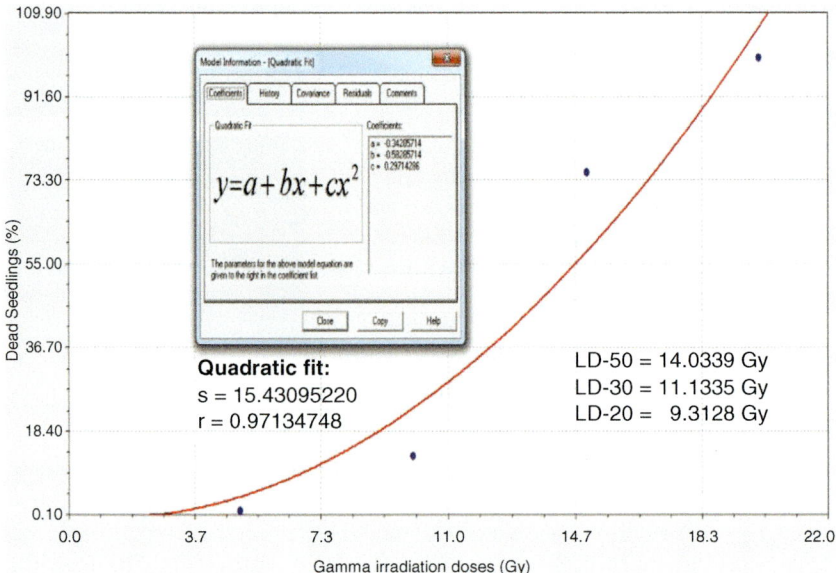

Fig. 3.2. Graph of lethality against gamma irradiation treatments.

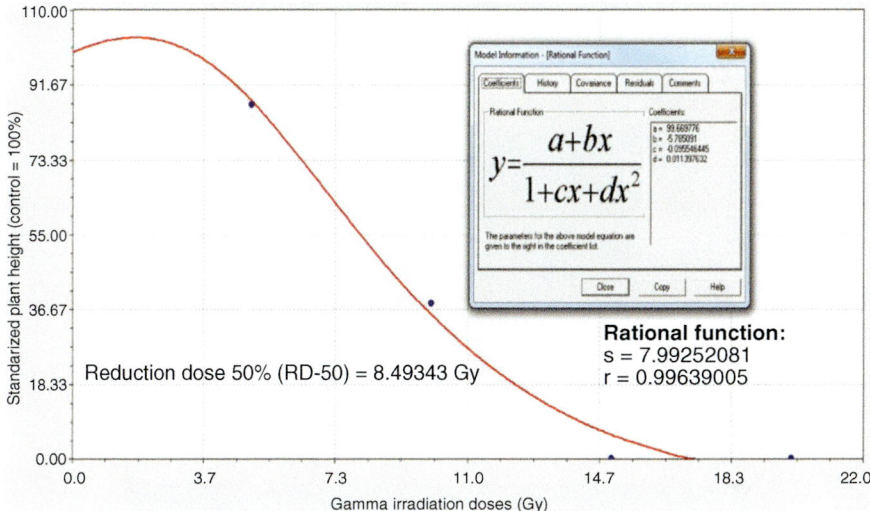

Fig. 3.3. Graph of reduction doses with the radiation rate.

In radio-sensitivity studies of reduced growth, the RD value is usually measured from the reduction of plant height, seedling height or leaf growth of M_1 plants compared to M_0 controls (no treatment). The best-fitting software analysis found that responses of gamma irradiation doses to seedling growth reduction in oil palm followed the rational function model given in Fig. 3.3. From this function, it was found that the 50% reduction dose, or RD_{50}, was equal to 8.49 Gy, which meant that, at that dose, the seedling height was reduced by 50%.

It should be noted that the LD data give slightly different results than the RD data; for example, LD_{50} is produced by 14.0 Gy, whereas RD_{50} is produced from 8.5 Gy. Thus, suitable irradiation doses for mutation induction range from 8 to 14 Gy.

3.5 Other Considerations

It should also be noted that the time in the gamma irradiator to give a specific dose treatment is dependent on the radioactive activity of the source (this declines with time). The time in a new irradiator (or newly refurbished irradiator) will be much less than in one where the half-life of the radioactive material has been reached. The activity of the irradiator, regardless of the dose given, is a major factor in mutation induction, and a note of the gamma source activity should be made. The experiments above were carried out using a digitalized Gammacell 220 available at the National Nuclear Energy Agency of Indonesia (BATAN, Java), as shown in Figs 3.4 and 3.5. The safety operation of the Gammacell follows BATAN's standard operational procedure No.

Fig. 3.4. Digitalized Gammacell 220 available at BATAN, Java, Indonesia.

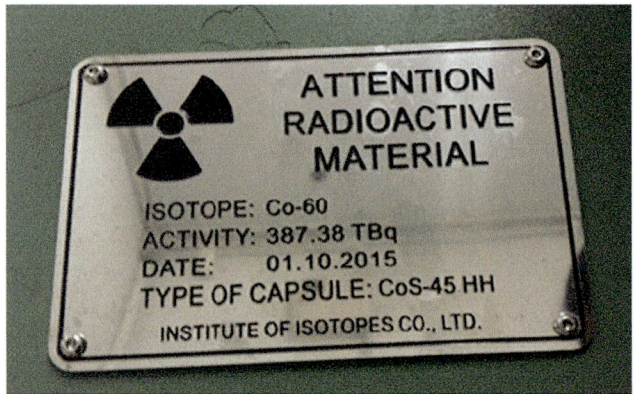

Fig. 3.5. Description of isotope source and activity.

008.003/IN0006AIR.8, referred to in the IAEA (International Atomic Energy Agency) standard safety recommendations for working with irradiation.

3.6   Steps in Irradiation Services at BATAN

The radio-sensitivity tests described above were carried out in collaboration with the National Nuclear Energy Agency (BATAN), Java, Indonesia, and involved the following steps.

Step 1

Sufficient germinated oil palm seeds are prepared (see Section 3.2 and Table 3.1) and delivered to BATAN. Prior to treatment, the seed is referred to as the M_0 generation.

Step 2

Prior to irradiation, the customer is required to fill in a registration form giving information on: name; date; kind of breeding material; size; number of samples; and required dose treatments.

Step 3

A professional technician will then irradiate the breeding materials according to the required dose treatments. Once the seeds have been placed in the machine, the irradiation dose has been set up at the control panel and the start button is pressed, the irradiation process runs automatically for a certain period, depending on the given dose and the dose rate at the time.

Step 4

The irradiated germinated seed (M_0) is returned to the breeder for immediate sowing in benign conditions, such as in small polybags in a shaded and irrigated nursery.

Only competent persons are allowed to enter the gamma room and they should be equipped with safety measures such as a thermoluminescent dosimeter (TLD). To maintain the accuracy of the gamma dosimeter, it should always be calibrated with a standard Fricke dosimeter every 3 months. The Fricke, or ferrous sulfate dosimeter, is a very useful chemical system for gamma dose measurements. Under irradiation, the Fe^{2+} ions are oxidized to Fe^{3+} ions, the concentration of which is proportional to the absorbed dose in the solution. The concentration of the Fe^{3+} ions is usually determined by measuring the optical density of the solution with a spectrophotometer at a wavelength of 304 nm. The standard Fricke dosimeter can be used in the dose range 4000–40,000 rads for gamma radiation. Since it provides an accurate and direct dose determination, it can be used to calibrate other systems, for example, ionization chambers.

References

Ghanim, M.A.G., Spencer Lopez, M. and Thomas, W.T.B. (2018) Mutation breeding in seed propagated crops. In: Spencer Lopez, M., Forster, B.P. and Jankuloski, L. (eds) *Manual of Plant Mutation Breeding*, 3rd edn. FAO/IAEA publication, FAO, Rome (in press).

Kelanaputra, E.S., Nelson, S.P.C., Setiawati, U., Sitepu, B., Nur, F. et al. (2018) *Seed Production in Oil Palm: A Manual*. CAB International, Wallingford, UK.

Kodym, A., Afza, R., Forster, B.P., Ukai, Y., Nakagawa, H. and Mba, C. (2012) Methodology for physical and chemical mutagenic treatments. In:

Shu, Q.Y., Forster, B.P. and Nakagawa, H. (eds) *Plant Mutation Breeding and Biotechnology.* CAB International and FAO/IAEA, Wallingford, UK, Chapter 14, pp.169–180.

Mba, C., Afza, R. and Jain, S.M. (2010) Induced mutagenesis in plants using physical and chemical agents. *Plant Cell Culture. Essential Methods,* DOI: 10.1002/9780470686522.ch7.

Ndofunsu, D., Ndiku, L., Otono, B., Nakweti, K., Mba, C. and Till, B.J. (2015) *In vitro* gamma radiosensitivity test in Congolese cassava, *Manihot esculenta* Crantz accession. *Academia Journal of Biotechnology* 3, 001–005.

Owoseni, O., Okwaro, H., Afza, R., Bado, S., Dixon, A. and Mba, C. (2007) Radiosensitivity and *in vitro* mutagenesis in African accessions of cassava, *Manihot esculenta* Crantz. *Plant Mutation Reports* 1, 32–36.

Rohani, O., Samsul Kamal, R., Rajinder, S. and Mohd-Nazir, B. (2012) Mutation induction using gamma irradiation on oil palm (*Elaeis guineensis* Jacq.) cultures. *Journal of Oil Palm Research* 24, 1448–1458.

Tshilenge-Lukanda, L., Funny-Biola, C., Tshiyoyi-Mpunga, A., Mudibu, J., Ngoie-Lubwika, M. *et al.* (2012) Radio-sensitivity of some groundnut (*Arachis hypogaea* L.) genotypes to gamma irradiation: indices for use as improvement. *British Biotechnology Journal* 3, 169–178.

van Harten, A.M. (1998) *Mutation Breeding: Theory and Practical Applications.* Cambridge University Press, Cambridge.

Options for Mutation Breeding in Oil Palm

4

Abstract

Theoretical and practical options for mutation breeding in oil palm are discussed. In theory, haploids are considered the ideal targets, as the induced mutation can be fixed instantly on conversion to doubled haploids. However, haploid/doubled haploid technology is in its infancy in oil palm and therefore other practical targets need to be considered. The two obvious targets are pollen and seed, as these are produced in large numbers in breeding and commercial seed production. Schemes for pollen and seed irradiation and subsequent mutant population development are compared. The irradiation of germinated seed is currently considered to be the better approach in terms of convenience and time. A major constraint in mutation breeding in oil palm is the long life cycle. Oil palm has a long juvenile stage, and it takes 4–5 years from sowing a seed to getting seed of the next generation. Traditionally, mutant selection has relied on phenotypic selection, which can only take place in the second mutant generation (M_2) due to the presence of physiological disorders and chimeras in the M_1. However, now that the oil palm genome has been sequenced, it is feasible to select for mutants genotypically in the M_1. Early detection of mutants is extremely valuable in oil palm, as it saves space in growing up only selected mutants in the field. Mutation breeding is a non-GM technique that has been used for over nine decades. New methods in genetic manipulation include gene editing, which has huge potential for the future but is currently still under debate as to whether it is considered a GM method and is subject to restrictive regulations.

4.1 Challenges and Opportunities for Mutation Breeding in Oil Palm

As mentioned in Chapter 1 of this manual, there are challenges and opportunities for mutation breeding in oil palm. However, of prime concern is the time taken to develop and select mutants of interest. The easiest crops to work with in this respect are annual, inbreeding diploid species that are seed propagated. Oil palm ticks two of these four boxes: it is seed propagated and diploid (however, recent research shows that much of the oil palm genome is duplicated, indicating that oil palm is an ancient tetraploid; Singh *et al.*, 2013). It has been general practice to induce mutation in sessile seeds (Bado *et al.*, 2015) and to screen for mutant phenotypes, at the earliest, in the M_2 generation. This is because: (i) the M_1 generation suffers from physiological disorders as a result of the mutation treatment (physical or chemical) and requires benign growing conditions to maximize the production of the next generation; (ii) M_1 plants are chimeric, and not all mutations will be carried forward into germline cells, and thereby the next generation; and (iii) the vast majority of mutants are recessive and will not express a phenotype until the mutant locus is homozygous. Thus, the M_2 is normally the first opportunity to produce homozygous mutants (ideally, the M_2 is produced via selfing). Oil palm is a perennial plant and it takes up to 5 years to produce seed of an M_2 population (via selfing of M_1 palms), and it takes a further 5 years to assess mature M_2 plants for yield characters. Thus, whereas an M_2 population can be produced and fully phenotyped in less than a year in annual crops such as rice, wheat and barley (using rapid cycling techniques; Forster *et al.*, 2015), this process may take up to 10 years in oil palm. Another major issue for mutation breeding in oil palm is the space required to grow the M_2 population, which may be composed of a few hundred individual plants. The area required is two to three orders of magnitude greater than that for annual crops, and therefore a major constraint. Mutation breeding is therefore not attractive unless savings in time and space (and thereby costs) can be made.

The genome of oil palm has been sequenced (Singh *et al.*, 2013), and this can be exploited in the early selection of mutants using genetic markers. Thus, the M_1 population can be screened for mutation in target genes, such as those controlling yield, quality, agronomy, height, disease and pest resistance and tolerance to abiotic stresses. Such genotypic selection can be performed on M_1 seedlings in the nursery, and selected plants transferred to the field for subsequent selfing to produce the M_2 population, which will exhibit novel phenotypes that may be selected for breeding. This strategy can save space and time in identifying valuable mutants at an early stage.

4.2   Options for Mutation Breeding in Oil Palm

Seed is an obvious target for mutation induction in oil palm, but oil palm pollen may also be considered as routine methods have been developed for pollen collection, storage and crossing in oil palm (Setiawati *et al.*, 2018).

Shell thickness in oil palm is controlled by the gene, *Sh* (Singh *et al.*, 2014). In commercial thin-shelled Tenera (*Sh/sh*) seed production, thick-shelled Dura (*Sh/Sh*) females are pollinated with shell-less and normally sterile female Pisifera (*sh/sh*) pollen. Pollen from elite Pisifera palms is therefore another potential target for mutation induction, as this is routinely collected by breeders and commercial seed producers. A scheme for mutation induction targeting pollen is given below.

Option 1 pollen irradiation: scheme, comments and timelines in producing mutant lines and populations in oil palm

Scheme	Comments and time considerations.
Irradiation of Pisifera pollen	It is normal practice to select elite material, which is readily available; therefore, Pisifera pollen is an obvious choice.
↓	
Cross on to best Dura	The irradiated pollen should be crossed on to Dura types and not Tenera types, as Teneras will generate 50% female sterile Pisifera in the M_1; thus, half the population is of little value. It takes 5–6 months from pollination to seed production.
↓	
Tenera M_1 progeny	May be screened genotypically in the nursery.
↓	
Self	It takes 5 years from seed production to collect pollen and artificially self-pollinate. It takes a further 1 year for seed production of the M_2 generation.
↓	
M_2 generation	May be screened phenotypically as seed, during seed germination, nursery plants and mature field-grown plants.

A serious drawback for the pollen option is the time taken to produce homozygous (mainly recessive) mutant alleles. It takes 6 years to produce mature M_2 palms, and therefore phenotypic selection of mutant traits cannot begin until year 7 (nursery traits), and mature trait screening cannot begin until year 10.

Option 2 seed irradiation: scheme, comments and timelines in producing mutant lines and populations in oil palm

Scheme	Comments and time considerations.
Irradiate germinated Tenera seed	Top commercial Tenera cultivars selected as M_0.
↓	
Sow M_1 seed in nursery ↓	Genotypic selection in the nursery.
Transfer selected plants to field	Normal spacing to ensure good seed production.
Transfer random plants to field	To be used as a source of mutations in other traits. Initially, plant at high density, but may be thinned out to maximize the yield of the next generation of selected palms.
↓	
Seed production of M_2 population	The M_2 is the first opportunity to produce homozygous mutants and thus apply phenotypic selection of kernel, germination, seedling, juvenile and adult traits.

The seed irradiation option outlined above takes 5 years to produce mature M_2 palms. This is shorter than the pollen irradiation option, which is delayed by an additional seed production phase.

4.3 Haploid Irradiation

The pollen and seed irradiation options are available now for oil palm. However, there have been new developments in haploid and doubled haploid production in oil palm (Dunwell *et al.*, 2010). Haploid and doubled haploid biotechnology of oil palm is still in its infancy, but haploids are extremely attractive targets for mutation induction, especially if methods for conversion to doubled haploids are available as this would provide homozygous mutant lines instantly.

4.4 Gene Editing

Gene editing is an exciting new development that has huge potential for crop improvement. As opposed to physical or chemical mutagenesis, which is random, gene editing is targeted to a specific DNA sequence of the genome. The methods involve the use of enzymes that alter DNA

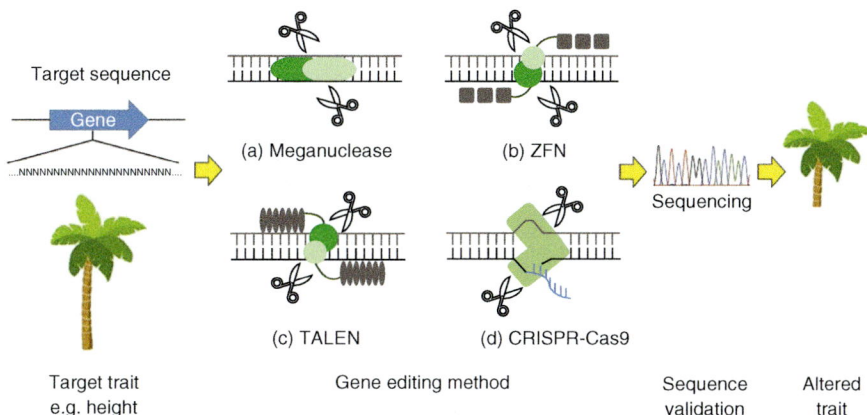

Target sequence

Gene

....NNNNNNNNNNNNNNNNNNNNNN....

(a) Meganuclease   (b) ZFN

Sequencing

(c) TALEN   (d) CRISPR-Cas9

Target trait   Gene editing method   Sequence   Altered
e.g. height   validation   trait

Fig. 4.1. Basic components in gene editing, starting with the selection of a target gene/trait (height is used as an example), to deployment of gene editing (four common methods are indicated), to the validation of edited gene sequence, to altered trait (from tall to short stature).

sequences: meganucleases; zinc-finger nucleases (ZFNs); transcription activator-like effector nucleases (TALENs); and clustered, regularly interspaced, short palindromic repeats (CRISPR-Cas9). The various enzymes have different DNA binding sites and cut the DNA (Abdallah *et al.*, 2015) and, importantly, are guided to the target gene sequence using various chaperones, for example Cas9; the target sequence is then replaced (Khatodia *et al.*, 2016). After the desired edits are achieved, the chaperones used can be removed. The basic components of gene editing are given in Fig. 4.1.

Gene editing has been applied successfully to major crops, wheat, rice, maize, soybean, tobacco and sweet orange (for reviews, see Baltes and Voytas, 2015, and Kumar and Jain, 2015). Currently, gene editing is regarded as a GM technique in many countries, as it involves recombinant DNA technology. This is a major concern for the oil palm industry that struggles to have an 'eco-friendly' and non-controversial (non-GM) image. The use of gene editing is therefore currently confined to R&D projects in oil palm. However, with the oil palm genome now sequenced (Singh *et al.*, 2013) and with obvious targets in oil palm, for example genes controlling oil quality, and with the deregulation of the technique, there is an exciting future for gene editing for crop improvement in oil palm.

However, until that time, the gap may be filled by traditional, non-GM mutation breeding techniques.

References

Abdallah, N.A., Prakash, C.S. and McHughen, A.G. (2015) Genome editing for crop improvement: challenges and opportunities. *GM Crops & Food* 6, 183–205.

Bado, S., Forster, B.P., Nielen, S., Ali, A.M., Lagoda, P.J.L. *et al.* (2015) *Plant Mutation Breeding: Current Progress and Future Assessment. Plant Breeding Reviews* 39. (Janick, J. (ed.)). Wiley-Blackwell, Hoboken, New Jersey, Chapter 2, pp. 23–88.

Baltes, N.J. and Voytas, D.F. (2015) Enabling plant synthetic biology through genome engineering. *Trends in Biotechnology* 33(2), 120–131.

Dunwell, J.M., Wilkinson, M.K., Nelson, S.P.C., Wening, S., Sitorus, A.C. *et al.* (2010) Production of haploids and doubled haploids in oil palm. *BMC Plant Biology* 10, 218–243. Available at: http://www.biomedcentral.com/1471-2229/10/218 (accessed 16 March 2018).

Forster, B.P., Till, B.J., Ghanim, A.M.A., Huynh, H.O.A., Burstmayr, H. and Caligari, P.D.S. (2015) Accelerated plant breeding. *CAB Reviews* No 43, 1–16. Available at: http://dx.doi.org/10.1079/PAVSNNR20149043 (accessed 16 March 2018).

Khatodia, S., Bhatotia, K., Passricha, N., Khurana, S.M.P. and Tuteja, N. (2016) The CRISPR/Cas genome-editing tool: application in improvement of crops. *Frontiers in Plant Science* 7, 506.

Kumar, V. and Jain, M. (2015) The CRISPR-Cas system for plant genome editing: advances and opportunities. *Journal of Experimental Botany* 66(1), 47–57.

Setiawati, U., Sitepu, B., Nur, F., Forster, B.P. and Dery, S. (2018) *Crossing in Oil Palm: A Manual.* CAB International, Wallingford, UK.

Singh, R., Ong-Abdullah, M., Low, E.T.L., Manaf, M.A.A., Rosli, R. *et al.* (2013) Oil palm genome sequence reveals divergence of interfertile species in old and new worlds. *Nature* 500, 335–341.

Singh, R., Ti, L.L.E., Ooi, C.L.L., Ong-Abdullah, M., Manaf, M.A.A. *et al.* (2014) *SureSawit*™ *Shell – A Diagnostic Assay to Predict Oil Palm Fruit Forms.* MPOB (Malaysian Palm Oil Board) TT No 548 (June), 656.

Protocol for Developing Mutant Generations for Mutant Selection

5

Abstract

A practical step-by-step protocol is presented for mutation induction and mutation detection in oil palm. Germinated seed is chosen as the target material for mutation induction, as this provides the quickest development of mutant populations (as discussed in Chapters 3 and 4 of this manual). The protocol adopts gamma irradiation as this is a proven effective mutagen for mutation breeding. Genotyping is deployed in the M_1 to select for mutants in target genes. These selections, plus a random selection of M_1 plants, are then advanced from nursery to field conditions to produce mature palms, which may be self-pollinated to produce the M_2 generation. The M_2 generation is subject to phenotypic screening at all stages in plant development, from seed, germination, seedling, juvenile to adult palms. A list of target genes and traits for mutation is given.

As discussed in Chapter 4 of this manual, there are two practical targets for mutation induction in oil palm: pollen and seed. The seed option is the more favoured, as it takes less time and involves only one pollination/seed production stage, thus saving time and labour. A step-by-step guide is provided in generating the M_1 and M_2 populations for mutant detection.

5.1 Step 1: Selection of Target Material (M_0)

Since the objective of the breeder is to produce a new cultivar as soon as possible, the aim is to target elite germplasm or top-performing cultivars that may be improved by the introduction of a favourable mutation for a trait of interest, i.e. with the minimum breeding effort. The obvious choice is commercial Tenera material, as this not only represents the best current material but also, since it is produced from a Dura × Pisifera cross, is highly heterozygous. The advantage in targeting heterozygous material is that mutation induction may knock out a dominant allele at a heterozygous locus and thereby expose a recessive phenotype at an early stage.

© Fazrin Nur, Brian P. Forster, Samuel A. Osei, Samuel Amiteye, Jennifer Ciomas, Soeranto Hoeman and Ljupcho Jankuloski 2018. *Mutation Breeding In Oil Palm: A Manual.*

The size of the M_0 population is determined by the available facilities of the investigator. Here, we used a population of 1000 as an exemplar.

5.2 Step 2: Germination of Target Seed (M_0)

Although the seed of Tenera palms is 'thin shelled', Teneras are produced mainly from Dura × Pisifera crosses, and since the shell is inherited maternally, the Tenera seed is encased in a thick shell. The thickness of the shell is a barrier for irradiation and can vary within and between genotypes. Also, oil palm seeds suffer from dormancy and viability, and thus germinated seeds are preferred as targets for irradiation. Germination of oil palm seed is a commercial process and seed may be bought pre-germinated. Methods in oil palm seed germination are given in Kelanaputra *et al.* (2018). The germination rate of the selected material needs to be considered to provide 1000 germinated seeds; commercial germination rates are normally above 80%.

5.3 Step 3: Seed Irradiation

Prior to seed irradiation for mutation induction, a radio-sensitivity test should be performed to determine the optimal dose treatment (see Chapter 3 of this manual). Normally, a dose treatment that delivers a 30–50% lethal dose (LD_{30-50}) is chosen. However, since oil palm produces hundreds of seeds from an inflorescence, a higher dose treatment is feasible. This will reduce the number of M_1 plants, but this may be compensated by the ability to produce relatively large M_2 populations from which desired mutants with a low background mutational load may be selected. In the scheme presented here, an LD_{50-60} treatment is used; for germinated seed at the early germination stage (Fig 5.1), the irradiation dose is 13–15 Gy, using a gamma cell

Fig. 5.1. Germinated seed at the right stage for irradiation treatment.

with an activity of 6363 Gy/h (Fig. 5.2). Thus, the M_0 seeds are exposed for 7.35–8.49 s. After irradiation, the seeds are referred to as the M_1 (first mutant) generation and the material is ready for immediate sowing. Therefore, it is important to be prepared to sow the material.

Fig. 5.2. A package of 100 germinated seeds is placed in the lift chamber and lowered into the gamma cell, where it is exposed to gamma irradiation (irradiation was performed at the National Nuclear Energy Agency (BATAN), Indonesia).

Irradiation must be carried out by a specialized institute, such as national or international atomic energy agencies: for example, Ghana Atomic Energy Commission (GAEC), Accra, Ghana; National Nuclear Energy Agency Indonesia (BATAN), Java, Indonesia; and the International Atomic Energy Agency (IAEA), Vienna, Austria.

5.4   Step 4: Sowing M_1 in the Nursery

The M_1 seedlings should be maintained stress free, as they are often weak, due to physiological changes, chimeras and mutational load caused by the irradiation treatment. The M_1 seeds should be sown into small polybags (23 cm deep × 15 cm diameter) and placed in an irrigated shade nursery (Fig. 5.3).

Fig. 5.3. Young M_1 seedlings in small polybags in a shaded and irrigated nursery (CSIR-OPRI, Ghana).

For an M_0 population of 1000 germinated seeds given an LD_{50-60} treatment, more than half the population is expected to die, leaving 400–500 M_1 plants for genotypic screening. This is a reasonable number for high-throughput DNA analysis.

After 3 months, the surviving seedlings are transplanted into large polybags (55 cm deep × 38 cm diameter; Fig. 5.4).

5.5   Step 5: Labelling of Surviving M_1 Nursery Seedlings

Since an LD_{50-60} dose was applied, 50–60% of M_1 plants were expected to die after a period of 3 months in the nursery. Therefore, genotypic mutation screening is not recommended before this time. Before screening, each surviving plant needs to be labelled individually (Fig. 5.5). In this scheme, 400–500 M_1 plants are expected to survive.

5.6   Step 6: Genotyping of M_1 Nursery Seedlings

Genotyping requires DNA extraction, DNA processing and DNA analysis techniques and sequence data of genes of interest. These procedures may be carried out in-house or outsourced. Some target genes for mutation detection are listed at the end of this chapter.

Fig. 5.4. Older (3–6 months) M_1 seedlings in large polybags in the main nursery at CSIR-OPRI, Ghana, in preparation for field planting.

Oil palm sequence data may be accessed from: https://www.ncbi.nlm. nih.gov/genome/?term=elaeis%20guineensis (accessed 19 March 2018).

For methods in DNA extraction, see Till *et al.* (2015). Alternatively, DNA kits (e.g. Qiagen) and sampling kits for specific genes (e.g. shell thickness, Orion) are available.

For various genotyping methods, see Pootakham *et al.* (2015).

5.7   Step 7: Selection of M_1 Plants for Field Planting

In annual crops, it is normal practice to grow up all viable M_1 plants and to produce selfed M_2 progeny of each fertile M_1 plant. This is unrealistic for perennial crops such as oil palm, for two reasons: (i) it takes 4–5 years from sowing a seed to getting seed of the next generation; and (ii) growing up plants to maturity takes a considerable amount of space. However, oil palm has an advantage in that it produces hundreds of seeds per female inflorescence, and therefore large mutant populations can be developed, providing a greater opportunity to find mutant plants of interest: thus, fewer M_1 plants are needed. In this scheme, M_1 plants are carried forward from both: (i) selected genotypic mutants; and (ii) a random sample.

Fig. 5.5. M$_1$ seedling in the nursery (Verdant Bioscience, Indonesia), labelled and bar coded.

Genotypic selections

Genotypic selections include all M$_1$ plants that exhibit mutations in genes of interest. With current knowledge, this may be restricted to a low number of target genes (10–50), with 5–10 plants being selected for each. However, as our knowledge of functional genomics increases, this will increase. Selected genotypic mutants should be grown in non-stressed conditions and at normal densities: 9 × 9 m spacing (143 palms/ha).

Random selection

It is recommended that a random selection of M_1 plants are taken forward to the field, as these provide novel mutation variation for future crop improvement. These would be given a lower priority than those selected for specific target genes, and so may be planted at a higher density to save field space. The number of random M_1 mutant plants taken will depend on the space available, but a minimum of 100 is recommended. In addition, any plants showing odd phenotypes should be selected, as some dominant alleles at heterozygous loci are expected to be knocked out, thus exposing recessive phenotypes (even though some of these odd phenotypes may be due to physiologic disorders rather than genetics). These selections may be planted at high density: 7 × 7 m (170 palms/ha). If some palms become more interesting, neighbouring palms may be culled to provide more space.

5.8   Step 8: Selfing of Selected Field-grown M_1 Palms to Produce M_2 Populations

It will take 4–5 years from sowing M_1 germinated seed to producing mature palms. Normally, the first inflorescences are male, and pollen should be collected from these and stored for subsequent self-pollinations when female inflorescences arise. Self-pollination is practised to produce homozygous mutant alleles in the M_2, thus revealing the phenotype of recessive alleles. Methods in pollen collection, pollen storage and pollination can be found in Setiawati *et al.* (2018). Methods in seed processing can be found in Kelanaputra *et al.* (2018).

5.9   Step 9: M_2 Phenotyping and Additional Genotyping

M_2 populations can be screened phenotypically at all growth stages from seed, seedlings, juvenile to mature palm trees and for harvest traits (Fig. 5.6 shows seedlings in the nursery, from which mutant phenotypes were identified, Fig. 5.7). The lack of efficient mass-screening methods can be a major bottleneck in selection, as desired mutants are expected at low frequencies. However, some efficient mass screens are available, for example, the observation of leaf colour mutants and plant stature in the nursery and seedling *Ganoderma* disease trials (Breton *et al.*, 2006; Rahmaningsih *et al.*, 2018). In addition, M_2 plants can be subject to further genotypic screening at all stages in development.

Fig. 5.6. M$_2$ seedling populations in a nursery (CSIR-OPRI, Ghana).

Broad-spectrum selection

In other species, notably small grain cereals, common mutations include pale green, short stature and late flowering. In barley, for example, it was found that these three mutant classes made up one-third of the mutant population, but importantly housed up to 55% of all other mutant pheno-types (Forster *et al.*, 2012). Therefore, by selecting pale green, short and late flowering M$_2$ plants, much of the mutant variation can be captured. For oil palm, it is not feasible to make early selection for fertility mutants, but the other two classes can be detected in the nursery and selected. It is also recommended that any visual mutant phenotypes observed in the nursery are also advanced to the field, provided there are no constraints on space. Depending on space, these may be planted at high density: 7 × 7 m (equates to 170 palms/ha). If some palms become more interesting, neighbouring palms may be culled to provide more space.

5.10   Target Genes and Traits

Current traits of interest to oil palm breeders are listed in Table 5.1. These may be screened for using genotypic and/or phenotypic tests.

Fig. 5.7. Examples of M_2 seedling mutant phenotypes (CSIR-OPRI, Ghana): (a) white streak; (b) green streak; (c) mottled; (d) pale green; (e) single leaf; (f) curly.

Table 5.1. Current traits of interest to oil palm breeders.

Genotypic screening	Phenotypic screening
Semi-dwarf	Semi-dwarf
Oil quality	Pale green
Shell thickness	Stalk length
Virescens	Frond length
Disease resistance	*Ganoderma* disease
	Fusarium wilt disease
	Drought tolerant
	Oil quality

References

Breton, F., Hasan, Y., Hariadi, Lubis, Z. and de Franqueville, H. (2006) Characterization of parameters for the development of an early screening test for basal stem rot tolerance in oil palm progenies. *Journal of Oil Palm Research*, Special Issue, 24–36.

Forster, B.P., Franckowiak, J.D., Lundqvist, U., Thomas, W.T.B., Leader, D. *et al.* (2012) Mutant phenotyping and pre-breeding in barley. In: Shu, Q.Y., Nakagawa, H. and Forster, B.P. (eds) *Plant Mutation Breeding and Biotechnology*. CAB International and FAO, Wallingford, UK, pp. 327–346. ISBN-19: 978-178064-085-3 (CABI); ISBN-13:978-925107-022-2 (FAO).

Kelanaputra, E.S., Nelson, S.P.C., Setiawati, U., Sitepu, B., Nur, F. *et al.* (2018) *Seed Production in Oil Palm: A Manual*. CAB International, Wallingford, UK.

Pootakham, W., Jomchai, N., Ruang-areerate, P., Shearman, J.R., Sonthirod, C. *et al.* (2015) Genome-wide SNP discovery and identification of QTL associated with agronomic traits in oil palm using genotyping-by-sequencing (GBS). *Genomics* 105, 288–295.

Rahmaningsih, M., Virdiana, I., Bahri, S., Anwar, Y., Forster, B.P. and Breton, F. (2018) *Nursery Screening for* Ganoderma *Response in Oil Palm Seedlings: A Manual*. CAB International, Wallingford, UK.

Setiawati, U., Sitepu, B., Nur, F., Forster, B.P. and Dery, S. (2018) *Crossing in Oil Palm: A Manual*. CAB International, Wallingford, UK.

Till, B.J., Jankowicz-Cieslak, J., Huynh, O.A., Beshir, M.M., Laport, R.G. and Hofinger, B.J. (2015) *Low-Cost Methods for Molecular Characterization of Mutant Plants*. Springer International Publishing. Available at: http://www.springer.com/gb/book/9783319162584 (accessed 16 March 2018).

Services in Irradiation Treatments 6

Abstract

Physical mutagens, particularly gamma and X-rays, are well established in plant mutation breeding. Although gamma irradiation has been predominant, it involves the use of radioactive isotopes and requires specialist facilities (gamma cells, gamma houses and gamma fields). X-ray machines are more abundant and easier to use, and so are becoming more popular as they do not involve radioactive isotopes and therefore are not governed by the same stringent regulations as gamma emitters. A brief description and comparison of gamma and X-ray irradiation is given. Physical irradiation services are available at the international, regional and national levels. Other non-physical mutagens are discussed in relation to plant breeding and functional genomics.

6.1 Physical Mutagens

Physical mutagens include ultraviolet light (a non-ionizing radiation) and several types of ionizing radiation: gamma, X-ray, alpha and beta particles, ion beam, ion implantation, protons and fast neutrons. Of these, gamma and X-ray are the most commonly used in plant mutagenesis, although new methods, particularly in the use of ion beam and ion implantation, are being developed. Details of the main physical mutagens used in plant mutagenesis are presented in Shu *et al.* (2012a,b), who provide information on their physical characteristics, mode of action and how they may be utilized in plant breeding and genetics.

Gamma rays

Gamma rays are emitted by the decay of unstable nuclei of atoms of radioactive isotopes. Compared to X-rays, they have a shorter wavelength but have greater energy per photon. Gamma irradiation is usually provided by isotopes of cobalt (^{60}Co) or caesium (^{137}Cs), and exposure is most commonly carried out in lead-shielded gamma cell irradiators. Gamma cells are particularly useful for acute (single dose) treatments of small samples; for example, for seed and small plant parts. Whole-plant treatments may be carried out in a gamma house or a gamma field, where chronic (prolonged or repeated) irradiation may be performed (Mba and Shu, 2012).

X-rays

Although gamma irradiation has been the predominant physical mutagen for plant mutagenesis over the past 4–5 decades, there have been major successes using X-rays (Mba *et al.*, 2012). Indeed, pioneering work on plant mutation used X-rays (Muller, 1927; Stadler, 1928). X-ray mutagenesis is currently experiencing a renaissance, as a range of X-ray machines are generally available (for example, in hospitals) and do not involve radioactive isotopes (which are subject to stringent regulatory controls). X-rays originate from electrons and not the nucleus of an atom. X-ray machines accelerate electrons in a vacuum, which are stopped abruptly by a barrier (e.g. tungsten or molybdenum). The collision results in radiation of various wavelengths (soft and hard X-rays). For mutation induction, hard X-rays (short wavelength) are usually preferred since they can penetrate plant tissues. Filters are used in hard X-ray production to remove unwanted soft radiation (Mehta and Parker, 2011).

6.2   Regulation of Irradiation Services

Radiation services are available at the international, regional and national levels. The International Atomic Energy Agency (IAEA) serves its Member States in the peaceful use of atomic technologies; this includes plant irradiation. The IAEA lists 169 Member States (see Table 6.1 below). These countries are scattered all over the world. Radiation facilities within Member States are controlled by their national energy agency, each of which is obliged to register their national facilities with the IAEA to obtain certification.

6.3   International Irradiation Services

The IAEA provides plant irradiation services through its Joint FAO/IAEA Plant Breeding and Genetics Laboratory, Seibersdorf, Austria. Gamma and X-ray irradiation services for plant mutation induction are available to its 170 Member States. Seed is the most common material sent for irradiation,

Table 6.1. IAEA member states (listed in April 2018).

Afghanistan	Djibouti	Lebanon	Rwanda
Albania	Dominica	Lesotho	Saint Vincent and the
Algeria	Dominican	Liberia	Grenadines
Angola	Republic	Libya	San Marino
Antigua and Barbuda	Ecuador	Liechtenstein	Saudi Arabia
Argentina	Egypt	Lithuania	Senegal
Armenia	El Salvador	Luxembourg	Serbia
Australia	Eritrea	Madagascar	Seychelles
Austria	Estonia	Malawi	Sierra Leone
Azerbaijan	Ethiopia	Malaysia	Singapore
Bahamas	Fiji	Mali	Slovakia
Bahrain	Finland	Malta	Slovenia
Bangladesh	France	Marshall Islands	South Africa
Barbados	Gabon	Mauritania	Spain
Belarus	Georgia	Mauritius	Sri Lanka
Belgium	Germany	Mexico	Sudan
Belize	Ghana	Monaco	Swaziland
Benin	Greece	Mongolia	Sweden
Bolivia, Plurinational	Grenada	Montenegro	Switzerland
State of	Guatemala	Morocco	Syrian Arab Republic
Bosnia and	Guyana	Mozambique	Tajikistan
Herzegovina	Haiti	Myanmar	Thailand
Botswana	Holy See	Namibia	The former Yugoslav
Brazil	Honduras	Nepal	Republic of Macedonia
Brunei Darussalam	Hungary	Netherlands	Togo
Bulgaria	Iceland	New Zealand	Trinidad and Tobago
Burkina Faso	India	Nicaragua	Tunisia
Burundi	Indonesia	Niger	Turkey
Cambodia	Iran, Islamic	Nigeria	Turkmenistan
Cameroon	Republic of	Norway	Uganda
Canada	Iraq	Oman	Ukraine
Central African	Ireland	Pakistan	United Arab Emirates
Republic	Israel	Palau	United Kingdom of Great
Chad	Italy	Panama	Britain and Northern
Chile	Jamaica	Papua New	Ireland
China	Japan	Guinea	United Republic of
Colombia	Jordan	Paraguay	Tanzania
Congo	Kazakhstan	Peru	United States of America
Costa Rica	Kenya	Philippines	Uruguay
Côte d'Ivoire	Korea,	Poland	Uzbekistan
Croatia	Republic of	Portugal	Vanuatu
Cuba	Kuwait	Qatar	Venezuela, Bolivarian
Cyprus	Kyrgyzstan	Republic of	Republic of
Czech Republic	Laos People's	Moldova	Viet Nam
Democratic Republic	Democratic	Romania	Yemen
of the Congo	Republic	Russian	Zambia
Denmark	Latvia	Federation	Zimbabwe

but other plant propagules (tubers, cuttings, etc.) and tissue culture samples may be sent. Procedures include:

1. Make contact with the Plant Breeding and Genetics Laboratory (contact details below) to discuss the service request with respect to the materials to be sent (and when), the irradiation service requested, treatment doses and the documents required.

2. High-quality plant materials should be selected and sent. These should be disease free, uniform and representative of the variety/line/genotype. The material should have a high level of vitality, for example, a high germination rate for seed, a high regeneration capacity for cultures (e.g. highly embryogenic cultures).

3. Plant materials (seed and plant part samples) should be bagged; tissue culture samples should be in sealed, sterile containers (e.g. screw cap tubes) and clearly labelled.

4. Some materials may require pretreatments. For example, sessile seeds are normally placed in a desiccator for some days to standardize the water content; dormant seed may require treatment to break dormancy; *in vitro* cuttings may need additional rounds of micropropagation to increase sample size.

5. The size (volume) of the samples should be determined prior to sending. The volume of the sample(s) is important, as the sample chamber for gamma cells may be a limiting factor.

6. If the required irradiation dose is not known, a radio-sensitivity test should be performed to determine the optimal treatment dose prior to irradiation of the main sample for mutation induction. Additional samples are required for this.

7. A Standard Material Transfer Agreement (SMTA) should be signed. The basic tenet here is that material will be returned to the customer and no third party will be involved.

8. Samples should be inspected by local quarantine officers prior to sending, and samples should be sent with a phytosanitary certificate.

9. Import quarantine permits and regulations will apply; be aware of these, they will vary from country to country and from sample to sample.

For more information on irradiation services, a service request form and further details, contact: Head, Plant Breeding and Genetics Laboratory, FAO/IAEA Laboratories, Reaktorstrasse 1, A-2444 Seibersdorf, Austria.

6.4 Regional Irradiation Services

An example of a regional (South-east Asia) irradiation service provider in an oil palm growing region is the Malaysian Nuclear Agency. The Malaysian Nuclear Agency offers a range of services, including: gamma cell irradiation; gamma greenhouse chronic irradiation; X-ray irradiation; and electron beam irradiation.

For information on irradiation services, contacts, procedures and request forms, visit the website: http://www.nuclearmalaysia.gov.my/new/PnS/services/radiationProcess/irradiationServices.php (accessed 19 March 2018).

6.5 National Irradiation Services

National or local irradiation services have several advantages as they are convenient (few transport issues) and, in most cases, preclude the need for quarantine.

Radiation facilities may be owned by the private sector, where supervision is maintained by the local national energy agency. In Indonesia, for example, there are private parties that have radiation facilities that remain under the supervision of the National Nuclear Energy Agency of Indonesia (BATAN) as the national energy agency. BATAN also functions as a collaborating centre with the IAEA.

For information on irradiation services contact: National Nuclear Energy Agency (BATAN), Centre for Application of Isotopes and Radiation, Jakarta 12710, Indonesia; website: https://www.batan.go.id (accessed 19 March 2018).

BATAN is a useful resource with respect to oil palm mutation breeding as it is located in the biggest oil palm producing country, Indonesia, in Southeast Asia. With respect to Africa, one may contact the Ghana Atomic Energy Commission (GAEC), who produced the first mutant oil palm population (currently grown at OPRI, Ghana). Contact details are: Ghana Atomic Energy Commission, PO Box 80, Legon, Accra, Ghana; website: https://gaecgh.org (accessed 19 March 2018).

6.6 Other Mutagens

In addition to physical mutagens, other mutagens include environmental, chemical and biological factors. Many spontaneous mutations can be attributed to environmental and biological agents. Although spontaneous mutants are rare, they can be significant in the domestication of crop plants from wild relatives, and in crop improvement. In oil palm, significant spontaneous mutations include the Pisifera shell-less fruit type allele, *sh* (Singh *et al.*, 2013) and the virescens allele, *Vir* (Singh *et al.*, 2015), which controls fruit ripening colour (see Chapter 1 of this manual). Plant stress can induce mutations through the activation of transposons and has been used to study gene function (Sangwan *et al.*, 2012; Zhu *et al.*, 2012). Transposons have been implicated in some significant crop traits, such as:

- reduced branching in the domestication of maize from teosinte (Studer *et al.*, 2011);
- red fruit flesh of blood oranges (Butelli *et al.*, 2012);

- oval shape of tomato fruit (Xiao *et al.*, 2008);
- white skin colour of grape fruit (Kobayashi *et al.*, 2004).

Chemical mutagenesis is used frequently in functional genome studies as it tends to produce a high mutational load (Leitão, 2012). Although this is good when you wish to produce mutations in as many genes as possible for functional genomics studies, for example for targeting induced local lesions in genomes (TILLING; McCallum *et al.*, 2000; Till *et al.*, 2012), there is a big disadvantage in plant breeding as chemical mutagenesis will destroy the genetic background of elite germplasm, which would then require a huge backcrossing programme to restore.

References

Butelli, E., Licciardello, C., Zhang, Y., Liu, J., Mackay, S. *et al.* (2012) Retrotransposons control fruit-specific, cold-dependent accumulation of anthocyanins in blood oranges. *The Plant Cell* 24(3), 1242–1255.

Kobayashi, S., Goto-Yamamoto, N. and Hirochika, H. (2004) Retrotransposon-induced mutations in grape skin color. *Science* 304(5673), 982–982.

Leitão, J.M. (2012) Chemical mutagenesis. In: Shu, Q.Y., Nakagawa, H. and Forster, B.P. (eds) *Plant Mutation Breeding and Biotechnology*. CAB International and FAO, Wallingford, UK, Chapter 12, pp. 135–158.

McCallum, C.M., Comai, L., Greene, E.A. and Heinikoff, S. (2000) Targeting induced local lesions IN genomes (TILLING) for plant functional genomics. *Plant Physiology* 123, 439–442.

Mba, C. and Shu, Q.Y. (2012) Gamma irradiation. In: Shu, Q.Y., Nakagawa, H. and Forster, B.P. (eds) *Plant Mutation Breeding and Biotechnology*. CAB International and FAO, Wallingford, UK, Chapter 8, pp. 91–98.

Mba, C., Afza, R. and Shu, Q.Y. (2012) Mutagenic radiation: X-rays ionizing particles and ultraviolet. In: Shu, Q.Y., Nakagawa, H. and Forster, B.P. (eds) *Plant Mutation Breeding and Biotechnology*. CAB International and FAO, Wallingford, UK, Chapter 7, pp. 83–90.

Mehta, K. and Parker, A. (2011) Characterization and dosimetry of a practical X-ray alternative to self-shielded gamma irradiators. *Radiation Physics and Chemistry* 80(1), 107–113.

Muller, H.J. (1927) Artificial transmutation of genes. *Science* 66, 120–122.

Sangwan, R.S., Ochatt, S., Nava-Saucedo, J.-E. and Sangwan-Norreel, B. (2012) T-DNA insertional mutagenesis. In: Shu, Q.Y., Nakagawa, H. and Forster, B.P. (eds) *Plant Mutation Breeding and Biotechnology*. CAB International and FAO, Wallingford, UK, Chapter 3, pp. 489–506.

Shu, Q.Y., Forster, B.P. and Nakagawa, H. (2012a) Principles and applications of plant mutation breeding. In: Shu, Q.Y., Nakagawa, H. and Forster, B.P. (eds) *Plant Mutation Breeding and Biotechnology*. CAB International and FAO, Wallingford, UK, Chapter 24, pp. 301–326.

Shu, Q.Y., Forster, B.P. and Nakagawa, H. (2012b) *Plant Mutation Breeding and Biotechnology*. CAB International and FAO, Wallingford, UK, 608 pp. ISBN-19: 978-178064-085-3 (CABI); ISBN-13:978-925107-022-2 (FAO).

Singh, R., Low, E.T.L., Ooi, L.C.L., Ong-Abdullah, M., Chin, T.N. *et al.* (2013) The oil palm *Shell* gene controls oil yield and encodes a homologue of SEEDSTICK. *Nature* 500, 340–344.

Singh, R., Ooi, L.C.C., Low, L.E.T., Ong-Abdullah, M., Nagappan, J. *et al.* (2015) *SureSawit™ Vir – A Diagnostic Assay to Predict Colour of Oil Palm Fruits.* MPOB (Malaysian Palm Oil Board) TT No 568, 697.

Stadler, L.J. (1928) Genetic effect of X-rays in maize. *Proceedings of the National Academy of Sciences of the United States of America* 14, 69–75.

Studer, A., Zhao, Q., Ross-Ibarra, J. and Doebley, J. (2011) Identification of a functional transposon insertion in the maize domestication gene tb1. *Nature Genetics* 43(11), 1160–1163.

Till, B.J., Zerr, T., Comai, L. and Henikoff, S. (2012) A protocol for TILLIING and Eco-TILLING. In: Shu, Q.Y., Nakagawa, H. and Forster, B.P. (eds) *Plant Mutation Breeding and Biotechnology.* CAB International and FAO, Wallingford, UK, Chapter 22, pp. 269–286.

Xiao, H., Jiang, N., Schaffner, E., Stockinger, E.J. and van der Knaap, E. (2008) A retrotransposon-mediated gene duplication underlies morphological variation of tomato fruit. *Science* 319(5869), 1527–1530.

Zhu, Q.H., Upadhyaya, N. and Helliwell, C. (2012) Transposon mutagenesis for function genomics. In: Shu, Q.Y., Nakagawa, H. and Forster, B.P. (eds) *Plant Mutation Breeding and Biotechnology.* CAB International and FAO, Wallingford, UK, Chapter 39, pp. 507–522.

Index

Page numbers in **bold** type refer to figures and tables.

CABI – who we are and what we do

This book is published by **CABI**, an international not-for-profit organisation that improves people's lives worldwide by providing information and applying scientific expertise to solve problems in agriculture and the environment.

CABI is also a global publisher producing key scientific publications, including world renowned databases, as well as compendia, books, ebooks and full text electronic resources. We publish content in a wide range of subject areas including: agriculture and crop science / animal and veterinary sciences / ecology and conservation / environmental science / horticulture and plant sciences / human health, food science and nutrition / international development / leisure and tourism.

The profits from CABI's publishing activities enable us to work with farming communities around the world, supporting them as they battle with poor soil, invasive species and pests and diseases, to improve their livelihoods and help provide food for an ever growing population.

CABI is an international intergovernmental organisation, and we gratefully acknowledge the core financial support from our member countries (and lead agencies) including:

 UKaid from the British people

Ministry of Agriculture People's Republic of China

 **Australian Government** **Australian Centre for International Agricultural Research**

 Agriculture and Agri-Food Canada

 Ministry of Foreign Affairs of the Netherlands

 Schweizerische Eidgenossenschaft Confédération suisse Confederazione Svizzera Confederaziun svizra **Swiss Agency for Development and Cooperation SDC**

Discover more

To read more about CABI's work, please visit: **www.cabi.org**

Browse our books at: **www.cabi.org/bookshop**, or explore our online products at: **www.cabi.org/publishing-products**

Interested in writing for CABI? Find our author guidelines here: **www.cabi.org/publishing-products/information-for-authors/**